Insect Control in the People's Republic of China

CPRC REPORT NO. 2

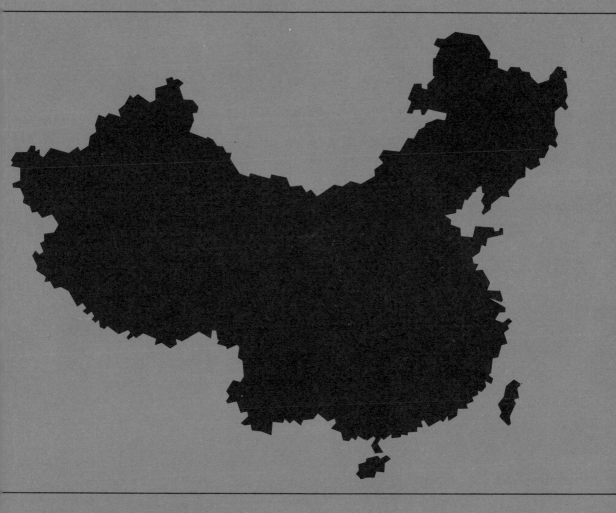

COMMITTEE ON SCHOLARLY COMMUNICATION
WITH THE PEOPLE'S REPUBLIC OF CHINA

CSCPRC REPORT NO. 2

Insect Control in the People's Republic of China

A Trip Report of the American Insect Control Delegation

Submitted to the Committee on Scholarly Communication
with the People's Republic of China

NATIONAL ACADEMY OF SCIENCES
Washington, D.C.   1977

NOTICE: The views expressed in this report are those of the members of
the American Insect Control Delegation and are in no way the official
views of the Committee or its sponsoring organizations--the American
Council of Learned Societies, the National Academy of Sciences, and the
Social Science Research Council.

Library of Congress Catalog Card Number 76-52849

International Standard Book Number 0-309-02525-7

Available from
Printing and Publishing Office
National Academy of Sciences
2101 Constitution Avenue
Washington, D. C. 20418

Printed in the United States of America

PREFACE

The American Insect Control Delegation, comprising nine scientists, a professor of Chinese history, and a staff member of the Committee on Scholarly Communication with the People's Republic of China, spent 4 weeks in China from August 3 to August 28, 1975. The delegation was organized under the auspices of the Committee on Scholarly Communication with the People's Republic of China (CSCPRC), jointly sponsored by the American Council of Learned Societies, the National Academy of Sciences, and the Social Science Research Council.

The decision on the part of the CSCPRC to propose an insect control delegation visit to China was taken after the Scientific and Technical Association of the People's Republic of China (STAPRC) sent an insect hormone group to the United States in June 1973. Several of the American scientists on the Insect Control Delegation were drawn from institutions visited by the Chinese in 1973. The American delegation was able to meet with all the available members of the Chinese Insect Hormone Delegation and to talk with them about the work that they were doing in the summer of 1975. There was evidence of an influence by the 1973 U.S. visit in ongoing Chinese projects, especially in the insect pheromone area.

The members of the Insect Control Delegation represented all aspects of entomology, with a strong focus on insect control and integrated pest management. Some of the group's more specific interests were in the areas of toxicology and pesticide development, pheromones and hormones, biological control, host-plant resistance, insect taxonomy, ecology, and physiology.

The economic entomologists were particularly interested in insects associated with cotton, soybeans, rice and other grains, citrus and deciduous fruits, and vegetables, and insects affecting man and animals. Examination of the biological control organisms that are associated with pest species affecting Chinese agriculture was a delegation goal. They also sought to observe the format and procedures involved in the program of insect population forecasting and the associated educational programs provided for farmers.

In the composition of the delegation's membership, an effort was made to include as broad a spectrum of specialists as possible within the general pest-control area. Persons representing research and professional interests in economic entomology in general, chemical control and the development of insecticides, medical entomology, insect ecology and population dynamics, insect physiology, insect pest management for field crops, biological control, and the development of crop resistance to insect pests were included. An effort was also made to include representatives of land-grant college teaching and research faculties, state extension programs, and government research programs from all parts of the country.

The delegation's host in China, the STAPRC, did much to make the visit a success. An excellent staff of coordinators and interpreters traveled with the delegation throughout the trip. Included were:

苏凤林   Su Feng-lin
STAPRC staff coordinator

苏全昌   Su Ch'uan-ch'ang
STAPRC interpreter

翟启慧   Chai Ch'i-hui
technical interpreter
(professor, Institute of Zoology)

邵明信   Shao Ming-hsin
STAPRC interpreter

Mr. Su Feng-lin was responsible for the arrangements made for the delegation to visit the laboratories, test plots, demonstration farms, and agricultural exhibitions of institutes, universities, provincial and municipal agricultural colleges and academies, communes, production brigades, and production teams in a wide selection of Chinese agricultural areas.

The STAPRC organized opportunities for delegation members to meet with scientists, technical specialists, factory specialists, farmers, and plant protection specialists. These meetings included formal exchange of prepared talks and lengthy discussions, which afforded the U.S. entomologists an excellent opportunity to learn about Chinese agriculture and pest control programs currently being used in China. Associated with this scientific itinerary was the chance to be briefed by, and to talk with, leaders of political and administrative units from the provincial level down to the production team. This gave the delegation opportunity to better understand the unique model of social and political organization that has produced outstanding agricultural achievements for China.

The STAPRC was especially helpful in making changes in the itinerary requested by delegation members, thus allowing the scientists to further explore avenues of interest as they developed throughout the trip. This flexibility was found not only in scientific interests but also in social and cultural interests.

Although the visit was limited, in both time and itinerary, the delegation felt that the cross section of integrated pest-management operations that they observed was broadly representative. Therefore, the report that has been produced is as fair an assessment as is available of China's national policies and practices in this vital aspect of the country's agricultural development policy.

This report is the joint effort of the whole delegation and reflects the general impressions of the group after their month in China. The report was completed with the help of Mr. Robert Neuman, Agricultural Editor of Michigan State University, and with partial support by Michigan State University.

MEMBERS OF THE AMERICAN INSECT

CONTROL DELEGATION

DR. GORDON E. GUYER (Chairman), Director, Cooperative Extension Service, Michigan State University, East Lansing, Michigan 48823

DR. PERRY L. ADKISSON, Chairman, Department of Entomology, Texas A & M University, College Station, Texas 77843

DR. HUAI C. CHIANG, Professor of Entomology, University of Minnesota, St. Paul, Minnesota 55108

MR. GALEN W. FOX, Foreign Service Officer, Department of State, Washington, D.C. 20520

DR. CARL B. HUFFAKER, Professor of Entomology, Director, International Center for Biological Control, University of California, Berkeley, 1050 San Pablo Avenue, Albany, California 94706

DR. FOWDEN G. MAXWELL, Coordinator, Environmental Quality Activities, Office of the Secretary, U.S. Department of Agriculture, Washington, D.C. 20250

DR. ROBERT L. METCALF (Vice Chairman), Professor of Entomology, University of Illinois at Urbana-Champaign, Urbana, Illinois 61801

DR. HAROLD T. REYNOLDS, Professor of Entomology, University of California, Riverside, Riverside, California 92502

DR. WENDELL L. ROELOFS, Associate Professor of Entomology, New York State Agricultural Experiment Station, Cornell University, Geneva, New York 14450

DR. BENJAMIN SCHWARTZ, Leroy Williams Professor of History and Government, Harvard University, Cambridge, Massachusetts 02138

DR. DONALD E. WEIDHAAS, Director, Insects Affecting Man Research Laboratory, Agricultural Research Service, U.S. Department of Agriculture, P.O. Box 14565, Gainesville, Florida 32604

MR. HALSEY L. BEEMER, JR. (Secretary), Staff Officer, Committee on Scholarly Communication with the People's Republic of China

Route followed by the American Insect Control Delegation in the People's Republic of China.

CONTENTS

1

GENERAL ASPECTS

A. IMPRESSIONS OF AN ENTOMOLOGICAL LAYMAN

When first informed of the opportunity to visit the People's Republic
of China as a member of the American Insect Control Delegation, I was
delighted. But I was apprehensive that as a nonentomologist and student
of Chinese history and politics, my experience in China might be even
more circumscribed and constricted than is generally the case on such
missions. In retrospect, however, it seems to me that the experiences
gained from our trip compare favorably in scope and variety with the experiences described in the accounts of many other delegations.

On the nonprofessional side, we had the opportunity to visit all the
usual sights. On the professional side, the area of insect control provided a focus through which one could glimpse many aspects of life.
Since the question of insect control largely involves the crucial areas
of agriculture and public health, we spent a considerable amount of time
in the countryside viewing the passing scene through car windows or
strolling in the fields and along rice paddy paths.

Since our entomologists were superbly competent in their scientific
specialties and thoroughly informed about every aspect of agriculture,
I had the privilege of observing the agricultural scene through their
eyes. It should be added that unlike some other areas of science, the
science of insect control is, in most of its aspects, accessible to the
interested layman. For the most part, I was able to follow the discussions.

The areas touched by the study of insect control included the whole
subject of science policy and science education, the organization of
scientific research, and the relationship between scientists and political cadres.

There were, to be sure, limitations. Since the central concern was
naturally insect control, we were unable to spend much time raising
questions on other matters in the communes and institutes that we visited. My own efforts to contact historians and intellectuals were not successful. Some account was taken, however, of my own interests as a
nonentomologist.

In Peking and Shanghai, I was allowed to detach myself from the
group occasionally for long strolls through the streets. At Peking University, I enjoyed the opportunity of perusing the stacks of the Chi-

nese section of the new library in the company of the Chief Librarian. At Changsha, we saw the exciting archaeological findings of the Ma Wangtui tombs. Yet most of my own impressions were derived from observations made in the course of our "official" business.

Our visit took place at a relatively quiet moment in the political history of the People's Republic. Chairman Mao's directive on studying the political experience of the Dictatorship of the Proletariat was a prevailing theme. But it did not seem that it had been allowed to become the subject of a frenzied campaign. The language of the "criticize Confucius, criticize Lin Piao" campaign was still on the lips of political cadres at all institutes and communal meeting places. But it was generally treated as a problem that had been resolved and "overcome."

It is a theme that continues, however, to lead a life of its own in the academic sphere where there has been considerable effort to rewrite the intellectual history of China in terms of the conflict between Confucianism and Legalism.

Before entering China and after leaving it, we heard stories of labor unrest at Hangchow and elsewhere. During the trip, we discerned no evidence of unrest. The debate about the novel *Water Margin* and the more recent campaign for the modernization of agriculture on the Ta Chai model had not yet surfaced. If the high-level conflicts that have since surfaced were already going on, such factional strife did not seem to have any observable impact on the scientific and agricultural activities we saw. On one occasion, I managed to ask the Chairman of the Tangwan Cotton Commune, in the vicinity of Shanghai, his views concerning the transition from the present system of communal organization to the level of "ownership by the whole people" (a topic frequently mentioned in the current ideological literature on the Dictatorship of the Proletariat). He made it clear that as far as he was concerned, this change was far off and that he was quite satisfied with the present state of affairs.

Our group was impressed with the achievements in agriculture. The policy of relative priority for agriculture (the policy of "agriculture as the foundation," which has prevailed on the whole since the early 1960's) seems to have produced remarkable results. There is maximum mobilization of labor power, and every bit of arable land (including new reclaimed terrace land in the Northwest) seems under intense cultivation.

In general, we found crops of high quality. While mechanization was evident in varying degrees, one was impressed by the degree of continued reliance on intensive labor. Tractors and trucks were seen in relative abundance in some areas, yet water buffalo and the hand-drawn cart continue to play a major role. The present discussion in China concerning increased mechanization would indicate a felt need for more rapid progress in this area. Nevertheless, remarkable achievements in water control, application of fertilizers, and use of electric pumps were noted everywhere.

Most of the vast hordes of people whom we saw in the cities and countryside seemed to be adequately nourished and clothed, although peasants still seem poorer than city folk. It appears that collectivization has been accepted, although the private plot and auxiliary family production still seem to play an important role.

We had almost no opportunity to make inquiry as to distribution systems in the various communes we visited. At the Tsaoyang Workers New Village in Shanghai and the Changchen Vegetable Commune, we visited households and interviewed housewives. Yet these were somewhat restricted experiences. While there seem to be serious questions about China's performance in agriculture, our overall impression, in terms of what was observed in the countryside, and judging from the appearance of most of the people, there has been considerable success in this vital area.

On the basis of my own observations and the judgments of members of our delegation, I have an overall impression of remarkable accomplishments in the realm of insect control. As one of our scientists has noted, "They are well versed in the current jargon of the integrated control specialists." The arguments put forth elsewhere for moving away from an exclusive reliance on insecticides would seem to be equally cogent for the Chinese. One heard arguments concerning harmful ecological effects and long-range ineffectiveness of insecticides as insects build up immunity. Even more than others, the Chinese seem to be particularly conscious of external costs of insecticide.

At the time of our visit, there seemed to be a tendency to rely on the expertise of scientists. One had the impression that the political authorities were relying heavily on the judgment of scientists concerning the scientific requirements of their discipline. The strategic role played by older scientists trained abroad or even in pre-1949 China was impressive. In Shanghai, Peking, Sian, and Kwangchow, we found that such persons seemed to be playing leading roles in planning, research, and implementing insect control programs on the provincial level. While many of these people had undergone vicissitudes during the Cultural Revolution, they now seemed to enjoy the confidence of the politically powerful. In addition to older scientists trained in the United States or in pre-Liberation China, a somewhat younger contingent trained in the U.S.S.R. in the early years of the People's Republic were also playing a crucial role.

It would be rash to draw conclusions from applied entomology to other sciences, such as theoretical physics, where the relationship between theory and practice is perhaps not as immediately obvious as in the case of the science of insect control. On the whole, however, there appeared to be considerable appreciation for the relevance of basic science, book learning, and reliance on international science, at least in the area of advanced research.

This does not mean that the Chinese claim of "self-reliance" is meaningless. There is still an enormous stress on making equipment and synthesizing insecticides in China. Some members of our group commented on the Chinese ability to do these things in austerely furnished laboratories. They contrasted this with the attitude of young American research scientists often unwilling to carry on unless provided with the most expensive and sophisticated equipment. In this sense, one can still speak of a reliance in China on men rather than equipment.

The Chinese also seemed more prepared than others to move forward rapidly in the application of new ideas, many of them developed abroad. Wide use of the *Trichogramma* wasp in the biological control of various noxious insects in the Kirin forests as well as in the South has, one gathers, never been implemented on this scale in the United States. The *Beauveria* fungus, which has been developed experimentally in the U.S.S.R. and the United States, is evidently being used to control corn borers at Ta Yushu brigade of Nan Weitzu Commune, Kirin Province. Certain inhibitions that exist in the United States against the widespread application of various techniques seem not to exist among the Chinese. The Chinese are not obliged to contend with the autonomy of the farmer, the fear concerning the possible noxious side effects of various fungi (such as *Beauveria*), and other governmental constraints involving possible health hazards.

Another aspect of insect control policy that seems to reflect the interplay of scientific and political realities is the degree of decentralization of science policy in this area. We observed a good deal of autonomy in the realm of insect control policy, particularly on the provincial level.

Genuine differences of opinion seemed to exist among these entomological experts (as they exist abroad) concerning the efficacy of various approaches and techniques, and the decentralization of decision making has led to different approaches in different areas. It was not easy to perceive the overall picture concerning the degree of communication and coordination among the institutes, universities, and communes that we visited. Thus, a considerable degree of autonomy for the scientific specialists was combined with a certain openness and flexibility in the application of policy on the local level.

There was, however, a question concerning scientific education. If the older generation of scientists who received a more intense type of training on more traditional lines was playing a leading role, was the present style of university science education producing a generation capable of taking their place? At Peking University, we were presented a detailed account of the new 3-year university program with its stress on work in the field and its strongly "applied" orientation. Entomology professors at Peking University, we found, tended to fall into a category quite separate from the research scientists. Their job, evidently, was teaching and not research.

One of the questions raised by some members of our delegation was whether the present 3-year academic programs provide an adequate grounding in basic science. If the universities continued to pursue their current methods of scientific training, would they produce the counterparts of the older foreign-trained scientists? It was in this area that one could discern the ongoing influence of the cultural revolutionary "radicals" who oppose in principle the image of the older types of scientific "specialists."

Yet, in observing some of the younger research scientists at institutes, I had the impression that the institutes themselves were providing a kind of advanced "graduate" training. They may to some extent fill the function of our graduate schools in the scientific sphere and thus fill the gap between the 3-year program and the demands of basic science.

The degree of flexibility and openness noted in the scientific sphere was not at all evident in any other sphere of culture. The 3-year university program in areas corresponding to our humanities and social sciences, as described to us in Peking, seemed with certain qualifications as constricted and ideology-bound as ever. There were no real signs of a return to the pre-1966 situation. One area where there is some possibility for creative work is archaeology, a science that continues to receive lavish government support. A good example of this was the impressive display of the Ma Wangtui tomb findings in the museum in Changsha. Another area where some work continues is intellectual history, where the Confucianism/Legalism discussion has provided a limited channel for the continuation of historic studies. It seems to provide the occasion for the republication of many classical works in the history of Chinese thought. It has, in effect, become the vehicle for the study of Chinese intellectual and social history.

The line to the past has not been sundered, although work in this area must be carried on within a severely constricted framework. There is more publication of books and periodicals; yet, in the end, the cultural policy of the People's Republic in areas other than science seemed frozen within very narrow limits (that is, narrow even in terms of the history of the People's Republic). This appeared true in academic "high culture" as well as in popular culture.

In communes and villages we visited, there seemed to be a total repression of popular religion in all its manifestations. The so-called art for the masses is frozen into the mold of the rococo formulas now favored as the people's art. It was my impression that the Maoist faith does not fill all the needs to which the popular religion and culture of the past addressed themselves (whatever the limits and "superstitious" nature of this religion and culture). I would say that while the pragmatic leadership of China seemed to be pursuing a policy of relaxation in the area of the natural sciences, it had not moved very far along these lines in other spheres of culture.

Our contacts with political cadres took place mainly at institutes, communes, and official banquets. While it was true as stated above that scientists enjoyed a degree of autonomy, ultimate authority was clearly in the hands of the politically responsible cadres (the chairman or vice-chairman of the institute's revolutionary committee).

The tone of the relationships between the political cadres and the local scientists differed markedly from place to place, depending on the personality of the dominant politically responsible persons. At one institute, the responsible person seemed overbearing and patronizing in his relationship to the research staff. But the politically responsible person at another institute seemed to enjoy a very open and trusting relationship with his staff. Whatever the style of leadership, scientists were being listened to on questions involving the requirements of their discipline.

We were not able to make many observations concerning the participation of the masses in political life. To be sure, we saw participation of the masses in the implementation of policies, but we had no opportunity to witness the degree to which the masses "are consulted." We were not able to attend meetings of production teams or factories where consultation with the masses presumably takes place. We saw the important role of mass involvement in public health measures and the propagation of educational devices used to convey pest control information. This, however, is not participation in decision making. We were in a much better position to observe the direction of the masses by the leaders. Here, there was a firmness of tone and a sure style of command. In the lower levels of leadership, most of the leaders had indeed come up through the ranks. They gave the impression of great ability, alertness, capacity for command, and willingness to command.

I came away with an impression of striking accomplishments in agriculture, public health, and insect control. While it is clear that the leadership is by no means completely satisfied with the present growth rates, what has been done, given the constraints, remains most impressive. During the period of our stay, it was clear that the political leadership had decided to make considerable concessions to scientific professionalism. In other areas of culture and education, however, there seemed to be little relaxation of an extremely constricted policy.

On the other hand, one did not have an impression of high ideological tension in the life of the people. I had the impression of a deep immersion on the part of most people in the course of their private and familial lives. If factional conflict was going on behind the scenes during our stay (as seems to have been the case), it had no discernible effect on these aspects of their lives.

Postscript

Since these observations were first set down, it seems that the winds of change have again begun to blow in the People's Republic. The death of Chou En-lai, the attacks on Teng Hsiao-ping, and the recent discussions of errors in the scientific and educational spheres may well have begun to affect some of the areas of science policy referred to above. We are now informed that a "right deviationist wind" was blowing in the scientific and educational spheres precisely during the period of our visit. Does this mean that the relative autonomy of the scientists and the prominent role of older scientists are again under review? Will the attack of the "radicals" remain largely verbal or will it affect policy? All of this again underlines the hazards of generalizations based on a short travel experience. The China of Mao Tsetung remains in flux.

by Benjamin Schwartz

B. STATUS OF AGRICULTURE

China is an agricultural country. Four-fifths of China's population resides in rural areas, where they are primarily engaged in what is still the most important task the Chinese rural people face: feeding themselves and city dwellers. The greatest achievement of the People's Republic has been to attain a basic self-sufficiency in grains even though its population is vast and growing, and only 11% of its land is arable.

China has reached basic self-sufficiency largely because its leadership has made a dedicated effort to do so. Mao Tsetung Thought, the official ideology of the People's Republic, proclaims the necessity to "take agriculture as the foundation." It calls upon the Chinese to eliminate the differences between rural and urban areas--a sophisticated indication of how serious China's leadership is about emphasizing agriculture. As the leaders realize, only if China's peasants are content to remain outside the urban areas can the agricultural system work to feed everyone. In fact, China has been remarkably successful in avoiding any vast, uncontrolled metamorphosis of peasants into urban unemployed, a process that has created many economic and social problems in developing countries.

Restrictions on travel, employment, government control of housing, rations cards, and the counterflow of youth and other city dwellers to the countryside--tools of coercion that can be employed by the leaders in an authoritarian and highly organized society like China's--keep the peasants where they are.

The diversion of monetary resources into agriculture that could have gone elsewhere, however, truly demonstrates how sincere the Chinese government is about developing its rural areas. Harvard economist Dwight Perkins estimates that the Chinese leadership's commitment to "take agriculture as the foundation" has been worth $5 billion annually to the agricultural sector since China shifted from its pre-1958 emphasis on rapid industrialization.[1] In the use of its precious hard-currency earnings, China has similarly favored agriculture. Among its most important purchases in recent years are 13 of the world's largest ammonia-urea fertilizer complexes.

The sincerity of China's commitment to agriculture seems well demonstrated. But this is part of the problem that Chinese agriculture faces today. Having committed itself so strongly to agricultural development in the past, China now finds it very difficult to further increase yields from its relatively small acreage of arable land. In each of China's major cropping areas, yields have already been pushed close to their natural limits:

● In the Northeast (soybeans, corn, sorghum), where the soil is rich and rainfall adequate, yearly yields are held down by the severe winters, which restrict the region to one crop a year. The best hope for increased output is expanding the acreage under cultivation by opening up more land in Heilungkiang Province, but there the bitter cold limits the annual growing season to 5 months.

● In the North China Plain (wheat), the chief problem is lack of water. In recent years, the People's Republic has undertaken a major effort to make the North self-sufficient in grain production. It built a vast system of 1.3 million tube wells combined with electric pumps capable of irrigating 7.3 million hectares of farmland by the end of 1974.[2] Yet this system failed to make the North self-sufficient, and it carries the danger of lowering the region's water table to a level where the wells run dry. Apparently the best long-term solution to North China's water problems is to reduce the silt content of the

Yellow River, which flows through the region, so water from the river can be used for irrigation.[3]

This, however, would involve a massive reforestation and dam construction project in the lightly populated areas of Northwest China where soil erosion into the Yellow River is most pronounced. It would be a very expensive effort.

● In Central and South China (rice), water is abundant and the mild climate permits double- and even triple-cropping. Arable land is heavily irrigated and fertilized and is already providing high per-hectare yields. Further increases are possible as China's ammonia-urea fertilizer complexes go into operation, but yields are not likely to rise in proportion to the added fertilizer.

New varieties of fertilizer-responsive seed in combination with increased fertilizer could raise yields significantly. But it was the conclusion of the CSCPRC Plant Studies Delegation, which visited China in 1974, that China would not develop such improved seed varieties unless the program of basic research in plant genetics was greatly improved.

Looming over China's difficulties in boosting its already high per-hectare yields is its population problem--5 million additional tons of grain must be raised each year just to cover China's annual population increase. This continuing necessity to increase grain production, along with the difficulties in doing so, ensures that agriculture will remain a fundamental concern of the Chinese leadership for years to come.

Within this context, control of pest damage to crops and the possible future development of insect-resistant strains of grains assume real significance, especially because the multiple-cropping and interplanting now widely used in China tend to complicate plant protection methods. Improving insect control is one key way in which China can increase grain yields.

In addition to providing its people with sufficient grains, China supplements the diet of its populace with fresh produce. Cities are fed from fruit, vegetable, and livestock communes located in nearby suburbs, under a distribution system that ensures a daily flow of fresh produce to district and local markets.

In the countryside, communes raise their own fresh produce. There is no food processing to speak of, and the fresh produce available within regions is subject to seasonal variations.

REFERENCES

1. Perkins, Dwight H. 1975. "Constraints Influencing China's Agricultural Performance," in the U.S. Congressional Joint Economic Committee's China: *A Reassessment of the Economy,* Washington, D.C., 1975, p. 365.
2. Erisman, Alva L. "China: Agriculture in the 1970's," in the U.S. Congressional Joint Economic Committee's China: *A Reassessment of the Economy,* Washington, D.C., 1975, p. 336.
3. Perkins, p. 361.

C. ENTOMOLOGICAL RESEARCH ORGANIZATIONS

Administrative Organization

Chinese agricultural research is divided into two bodies, both at ministry level and both reporting directly to the State Council. (Figure 1.) One body consists of research institutes administered by the Chinese Academy of Sciences (both directly and in conjunction with provincial and municipal governments). The other consists of research institutes and provincial academies (with their own research institutes) administered by the Ministry of Agriculture and Forestry through the Chinese

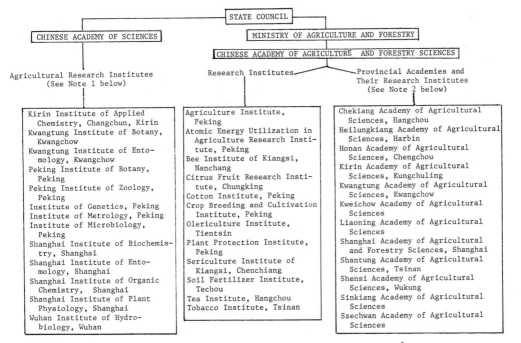

FIGURE 1 Organization of Chinese agricultural research[1]

Note 1 Some of the research institutes in the left-hand box are prefixed by a place name (e.g., *Peking* Institute of Zoology). These are administered jointly by the political unit mentioned in the title and by the Chinese Academy of Sciences. The other institutes are administered solely by the Chinese Academy of Sciences.

Note 2 The right-hand box lists provincial academies. The research institutes attached to these academies, and their locations, are as follows: Crop Breeding (Chekiang, Heilungkiang, Kirin, Shanghai, Shantung), Soil Fertilizer (Chekiang, Kirin, Kwangtung, Shanghai), Fruit Tree Research (Honan, Shantung), Plant Protection (Kirin, Kwangtung, Shanghai, Shantung), Grain Crops (Kwangtung), Peanuts (Shantung).

Information on the organization of agricultural science in China--here and in the text--is drawn from the Insect Control Delegation's information, from information gathered by the 1974 U.S. Plant Studies Delegation, and from U.S. Government sources.

Academy of Agricultural and Forestry Sciences. Generally, research institutes of the Chinese Academy of Sciences focus on basic research, while those of the Ministry of Agriculture and Forestry focus on applied research. The research functions of the Chinese Academy of Sciences and the Ministry of Agriculture and Forestry are tied with further networks of research groups that descend the administrative ladder below the provincial level. Although ties between the research units administered by the Chinese Academy of Sciences and lower units are not clear, those between units of the Ministry of Agriculture and Forestry are more easily discerned. For instance, the administrative structure for research in Kirin Province could be diagrammed as follows:

Administrative Level	*Unit Title*
National	Ministry of Agriculture and Forestry
	Chinese Academy of Agricultural and Forestry Sciences
Provincial	Kirin Academy of Agricultural Sciences
	Crop Breeding Institute
	Plant Protection Institute
	Soil Fertilizer Institute
District	District Agriculture and Forestry Office (seven district offices in Kirin)
County	County Scientific Experiment Station
Commune	Commune Scientific Experiment Station
Brigade	Production Brigade Scientific Experiment Station (29 basic points throughout the commune)
Team	Production Team Scientific Experiment Small-Group

An example of the Agricultural Extension System for Plant Protection, also in Kirin Province, will illustrate how the extension apparatus and the agricultural research apparatus are intertwined. The plant protection system for Kirin could be diagrammed as follows:

Administrative Level	*Unit Title*
Provincial	Kirin Province, Plant Protection Station
District	District Plant Protection Station (seven district stations in Kirin)
County	County Scientific Experiment Station

From the county level downward, the administrative network is identical to the research/experiment network.

Academies and Institutes

The Chinese Academy of Sciences (Academia Sinica) is a national, minis-try-level organization that coordinates research operations through its institutes in different parts of the country. Some institutes are di-rectly under the administration of the Academy and some are administered jointly by the Academy and by a provincial or municipal government.

Additionally, in major provinces various agricultural institutes were pooled in 1949 to form academies. These academies come under the administration of the Ministry of Agriculture and Forestry. (See chart on page 9.)

In the general administration of the research institutes of the Chinese Academy of Sciences and the Ministry of Agriculture and Forest-ry, a pattern can be discerned. The director of each institute is often a political person, having risen through the ranks of the party and pos-sessing some, but often limited, scientific background. The vice-director is often a scientist, usually a senior scientist with both research and teaching experience.

Specific Institutions Active in Entomological Research

In order to gain a general view of the organizations currently active in entomological research, 9 recent issues of *Acta Entomologica Sinica* were scanned.

The governmental organizations at various levels are as follows:

Level	*Institutions*	*Place*
National	Peking Institute of Zoology, Academia Sinica	Peking
	Chinese Academy of Agriculture and Forestry Sciences	Peking
Provincial	Institute of:	
	Apiculture	Kiangsi
	Biology	Chinghai
	Entomology	Kwangchow
		Shanghai
	Forestry	Kirin
		Shensi
	Forestry and Soils	Liaoning
	Fruit Trees	Shensi
		Hopei
	Oil Crops and Tea	Chekiang
	Plant Protection	Kwangsi
	Plant Protection and Soils	Hopei
	Rice Pest Biological Control	Kwangtung
	Sericulture	Kiangsu
		Shantung
District	Agricultural Work Station	
	Agricultural Science Institute	

Level	*Institutions*	*Place*
County	Plant Protection Station Agricultural Bureau Biological Control Station Science and Technical Bureau Experiment Station	
Commune	Agricultural Science Experiment Station	
Brigade	Technique Station	

Seventeen provinces and 3 national municipalities have research organizations.

Some subdivisions in some institutes are so specialized that they may be *ad hoc* organizations, e.g., a hybrid heterosis utilization unit, an insect repellent unit, or a leaf beetle group. Interlevel cooperation is evident. There are reports resulting from cooperation between institute scientists, local technicians, and commune peasants.

The universities contributing reports are: Peking University, Nankai University, North China Agricultural College, Chungchin Medical College, Chekiang Agricultural University, Liaoning Agricultural College, Kwangtung Agricultural and Forestry College, Kweichow Medical College, Shensi Teachers University, Kwangtung Forestry College, and Northeastern Agricultural College.

Basic Approaches in Research Organization

While traveling through China, the delegation was often briefed by officials from different educational bodies, research organizations, and extension programs. From these briefings, which often stressed the same approach to research, it is possible to outline some of the basic approaches to Chinese research organization. It should be understood that this description represents the ideal or the model of research organization as given to the delegation, and it was not always possible to verify whether the system actually worked this way.

Self Reliance A solution to a problem is always sought locally first. Upward requests for recommendations and information are made only when necessary, and then one level at a time. Recommendations are then handed down, shortening bureaucratic red tape and minimizing turnaround time.

Interlevel Cooperation Brigade and production teams have study "points" (sites) for their work. Scientists from higher level research organizations may also work on these points to cooperate and help the lower-level workers. They may stay a few months, or 1 to 3 years, or until a problem is solved or a new system developed. The interaction between interlevel personnel is maximized by the "three in one concept" in which food, shelter, and work are shared. Such programs have resulted in publications with joint authorships reflecting the interlevel cooperation.

Integration of Research and Production This is achieved by coordinating:

- Leadership--consists of leading cadres, leading peasants, and technical personnel.
- Research emphasis--equally on investigation, production, and utilization of agricultural products.
- Control methods--through the phases of testing, demonstration, and extension.

This approach expedites the pathway of research, extension, and production. For example, a good new-crop variety can be increased and put in production in 2 years after being developed, and an effective method of insect control can be popularized and adopted on a commune, or even larger, basis in 1 year.

A case study of the development and administration of a successful control program--use of *Trichogramma* to control the European corn borer--follows:

Problem identification With the increase in plant density and fertilizer application in Kirin Province, corn borer infestation increased. Prior to 1970, corn borer control programs involved application of sprays and granular insecticides. Labor investment was high and the effective period was short. The peasants were not satisfied. Thus, new methods were needed.

Program development Encouraged by the success of *Trichogramma* on the sugarcane borer in Kwangtung and on the pine caterpillar in Kirin, the possible use of *Trichogramma* on the corn borer was considered and promoted by most of the cadres and peasants. Approval was then obtained from the top-level Party committee to initiate a massive experiment on the use of *Trichogramma* to control the corn borer. With the total support of the local Science and Technology Bureau and the local Agriculture Bureau, a task force was organized in Sanke-yusha People's Commune, Tung-Hua County, Kirin, consisting of commune members, leading cadres, and technicians. In 1971, about 70 hectares (1,100 mu) of land were allotted for the project.

The project involved mass rearing and field release of the parasite. Methodology was developed with local resources on a make-do basis, such as the use of natural caves for cold storage, homemade temperature cabinets, and modification of a meat grinder to egg extractor.

Party support Encouragement was offered to the technical personnel. Commune members and peasants were involved in the project, participating in discussion, field inspection, and making suggestions to solve problems.

Training sessions were organized. Plant protection technicians were taught how to rear and release the parasites. The latter involves determination of percentage of parasitism of silkworm eggs being released, the placement of the eggs on corn, and, later, evaluating parasitism of corn borer eggs. These persons formed the *Trichogramma* Team, which carried out the project.

Extension sessions were organized. This aspect involved preparing and showing movies on procedures and holding demonstration meetings. The trained technicians of the *Trichogramma* Team served as extension specialists.

Feed-back sessions were organized. The peasants' response to the program was relayed to technicians. Changes were made to meet their needs.

Technical improvements

● Rearing The rearing program had to produce wasps in sufficient quantities and with vigor for dispersal and oviposition. Improvements on procedures were continuously made. These included limiting the generations of laboratory propagation, starting cultures with new and vigorous stock, varying the maintenance temperature, and using honey and royal jelly as wasp food.

● Releasing Release of parasites had to coincide with borer oviposition. Thus, monitoring and forecasting of borer phenology were important. Checkpoints were set up in each brigade. Weather forecasting was also used to avoid storms and high winds during release. At one time, parasitized eggs were pinned on corn plants, but the pins made the plants unsuitable for cattle feed. A new method of placement without metal pins was then developed.

● Efficiency and economy of production One person could produce enough wasps for 13.3 hectares (200 mu), and one person could handle the release in 3.3 hectares (50 mu). The total cost was about $1.35 per hectare (0.175 Chinese yuan per mu).

As a result of the experience up to 1974, two documents were prepared: *Manual on* Trichogramma *Rearing Technique* and *Procedures for the Large-Scale Demonstration of Trichogramma to Control Corn Borer in Tung Hua Region in 1974.*

Expansion of the program As members of other communes and other counties learned about the effectiveness and economy of the program, they asked for authority to include their communes in the program.

The increase in area covered by the program in 4 years was as follows:

	1971	1972	1973	1974
Tunghwa County	1,100 mu	17,500 mu	17,500 mu	300,000 mu
Liuho County	-	-	15,286 mu	51,000 mu
Hailung County	-	-	-	3,240 mu
Total	1,100 mu	17,500 mu	32,786 mu	354,240 mu
	71 ha	890 ha	2,106 ha	22,751 ha

Result evaluation Parasitism, borer damage, and borer populations in release and control areas in Sanke-yusha Commune in 1971 - 1974 were as follows:

Year	% Egg Parasitism			No. Tunnels/100 Plants			No. Borers/100 Plants		
	Con-trol	Re-lease	Effect (%)	Con-trol	Re-lease	Effect (%)	Con-trol	Re-lease	Effect (%)
1971	--	--	--	348.4	153.2	-55.8	214.0	65.2	-69.6
1972	0.0	78.0	+78.0	171.0	63.1	-63.1	71.5	20.5	-71.3
1973	10.6	91.2	+80.6	304.5	120.5	-60.4	191.4	38.8	-79.7
1974	--	86.3	--	200.0	32.0	-84.0	72.0	12.0	-83.4
Mean	5.3	84.6	+79.3	255.8	92.0	-65.6	139.6	33.9	-76.0

Sources This review describes initiation, execution, expansion, and evaluation of a project of a substantial scale. It also glimpses the decision-making process and program organization. The review was prepared on the basis of a presentation by a scientist at the Kirin Plant Protection Institute, and two articles in the Acta Entomologica Sinica:

Seed and Plant Protection Station, Tung Hua Region, Kirin Province. 1975. The experience and realization of large scale control of European corn borer by using Trichogramma egg parasites. Acta Entomologica Sinica *18*: 7 to 9.

Agricultural Experiment Station, Sanke-Hushu People's Commune, Tung Hua County, Kirin Province. 1975. The control of European corn borer by using Trichogramma egg parasites. Acta Entomologica Sinica *18*: 10 to 16.

REFERENCE

1. The information for this organizational chart is drawn from the delegation's information, that of the 1974 U.S. Plant Studies Delegation, and U.S. Government sources.

D. AGRICULTURAL EXTENSION "POPULARIZATION" PROGRAMS

There is ample evidence in China that the positive results of agricultural research and demonstrations are rapidly translated into action programs. Extension activities are collectively called "popularization," which effectively characterizes the priority associated with preparation and dissemination of information for the benefit of the masses.

Research is primarily associated with applied, or problem-solving, issues and is characteristically implemented at the local level, which encourages direct relationship between research and applied educational programs. This interaction is further strengthened by the fact that most researchers and their university students involved with insect control have direct contact with specific insect problems at the farm level.

In general, popularization programs are loosely structured and varied in organizational format and activity, their nature depending on the location, the problem, and local needs. However, there are many examples of highly structured, long-standing forecasting projects that are impressive and responsive to specific insect-control problems.

The extension structure appears to consist of a three-level network that receives information from various agricultural-science institutes and local facilities. This network includes a technical committee, scientific and technical stations in each brigade, and scientific and technical groups at the team level. Within this framework, the techniques of pest control are organized in association with other agricultural activities. Examples of these interactions are: principles of land cultivation, cultural control of insect pests, appropriate choice of seed and planting procedures, pest prognosis, water management, and fertilization programs.

Prognostic procedures have been established for many of the insect pests, especially by groups in production brigades and production teams, where pest status in each field can be used to predict the appropriate time to apply controls. This has contributed to effective pesticide use and integration of chemicals with other control procedures. It is clear that the communes, and more specifically, production brigades and teams are focal points for control decisions and for interaction of control techniques.

There is evidence that insect forecasting is well organized and effective. The delegation had an opportunity to observe the forecasting procedure at the Kirin Academy of Agricultural Sciences in Kungchuling, Kirin Province. Here it was emphasized that the 4-level program of combining provincial, district, county, and commune activities was extremely important in the development of forecasting activities.

In Kirin Province there are 186 forecasting stations. These include scientific research institutes, professional forecasting teams, agricultural communes, and farm-level forecasting teams (Figure 2). The provincial forecasting center is located at the Academy, and all 186 stations report to this center. The center forecast output includes occurrence and abundance of the pest, population trends, and suggested control methods. The forecasting provided control information to 500 units included in the province, district, county, and communes.

FIGURE 2 Workers
determining insect
populations. This
activity is typical
of extension pro-
grams used to
inform farmers of
proper control
methods.

Information includes trends for the entire year and shifts in population over a 10-day period. This forecasting and popularization unit also provides published materials. It was pointed out that this program had been in operation for more than 20 years. We were able to review the standardized forms used to record and maintain the data. These records provide a basis for predictions and for related bionomic information studies.

Forecasting uses information collected over many years from many points, and relates the populations to climatic conditions, taking into account the impact of natural enemies. Treatments are based on regression analyses of the data.

An example of the intensity of extension activities at the commune level was of particular interest during a visit to the Tangwan Cotton Commune. This commune has an excellent plant-protection station, with evidence of a high priority on educational efforts.

We were told that there are 4 forecasting points in the Tangwan Commune, which has 380 plant-protection specialists involved in extension activities in the commune, trained at the county plant-protection station. Specialists from the various institutes often lecture at training sessions. Pesticide information for the commune often originates at the provincial or county level.

When a pest problem is of regional or national significance, the extension program is conducted on a larger scale. It includes demonstrations, participation by technicians from the various institutes, large national meetings, unified development of extension literature, and input by commodity associations. A large-scale program having to do with a new virus for control of the mulberry tussock moth is being prepared. In all silk-producing areas of China where the tussock moth is a pest, this technique will be implemented in conjunction with activities of the National Silkworm Association.

The Chinese Association of Agriculture comprises scientists, administrators, and general agriculturists who provide additional educational expertise at the national, provincial, and local levels. In an extensive interview with key representatives of the Association, it was pointed out that they have a major educational commitment emphasizing the "three in one principle": involvement of (1) scientific leaders, (2) technicians, and (3) the masses. They organize educational meetings, publish extension-type literature, and in general serve in a coordinating role in the preparation and dissemination of educational materials. It appears that the Association is unofficially responsible for many of the functions and programs handled by professional societies and trade associations in the United States.

A statement presented at the Shanghai Institute of Entomology clearly defines the priority placed on extension activities:

"Parallel to the research work, we pay much attention to popularizing scientific knowledge through short-term classes and other forms of training technical personnel for the productive units. Through practice it has been demonstrated that a persisting and open-door institution, combining research members with workers and peasants, speeds up the remodeling of the world outlook of

the research member, raises the level of scientific work, and fa-
cilitates the extension of scientific results as well."

E. TEACHING PROGRAMS

Chinese Education System in General

Grade school in China is compulsory for 5 years. Besides regular subject matter, group discussions are a part of the children's curriculum.
Through such discussions, students develop ability to speak in public,
ability to work with peers, awareness of the political and social system in which they live, and a concern for the needs of the society.
They may serve as the link between their parents and the society at
large on social issues.
  The middle school is 5 years in length. Training in a foreign language is required. Currently, English has replaced Russian as the
popular choice, and the delegation was told that a shortage of English
language teachers is developing.
  All middle school graduates are required to perform 2 or 3 years of
"national service" in the army, in a factory, or on a farm. After this,
some will be assigned a job, at the same post or elsewhere, and some
will be selected to go to college.
  Selection of college students is an elaborate process. The government decides the number of entering college students needed in each
field for a given year. After consultation with college administrators,
a certain number of students are assigned to a certain university. For
example, it may be decided that 100 entering students in entomology will
be needed for 1976, and 40 will be at Peking University, 20 at Chungshan University, and so on. The 100 students will be drawn from army
units, production units, and service units. Similarly, there are al-
locations to other disciplines. Thus, a certain commune may be asked
to select, for example, one entomology, one engineering, and one chemistry candidate.
  Upon the receipt of this list, young people in that commune who have
completed at least 3 years of service may apply for the discipline of
their choice. Through group discussion, candidates are selected on the
basis of political consciousness, interest in the subject, motivation,
attitude, and ultimate service to the candidate's unit. In one case,
the Evergreen Commune near Peking recommended over 40 students for college programs in 1975.
  In any given year, if no one in a particular unit is interested in an
entomology allocation, the allocation will revert to the government for
reassignment. Fields that require special talent, such as music and
art, do not have quotas. In these fields, rigorous screening discourages all except those with impeccable political credentials, extreme
confidence, and determination. If a person has a good work record in
his unit, he will not find it very difficult to be recommended by his
unit for the screening evaluation at a local level. This will be only
the first of a series of tests before being admitted to a music school.

The delegation was told that students are so keenly aware of their responsibility--to those who selected them and to the country--and so carefully screened that they never fail and never shift their interest during their college career.

Students so selected receive the salaries that they received at their old jobs, even though they may not go back to the same jobs after graduation.

Students in agriculture must spend a third of their time on a farm. For example, 35 students of Tsing Hua University (an institute of technology) who were studying agricultural machinery worked on the Evergreen Commune in 1975.

Entomology Teaching Programs Visited

Peking University Peking University, established in 1898, is the oldest university in Peking. It has 75 departments in agriculture, natural sciences, and liberal arts. Enrollment before the Cultural Revolution was about 2,000. At the time of our visit, there were 5,000 students, but 8,000 were expected in the fall of 1975, and 10,000 in 1977. There were about 200 foreign students.

The student body in entomology consisted of 9 second-year students and 17 first-year students. Forty students were expected to enter in the fall of 1975. The lack of third-year students is the result of the Cultural Revolution, which halted regular scheduled classes. The increasing enrollment reflects the recognition of the importance of the subject. The department has 17 faculty members who teach insect taxonomy, morphology and anatomy, physiology, ecology, biochemistry, and experimental techniques (microtechnique, photography, specimen preservation, etc.). Physiology covers toxicology, and ecology covers insect forecasting.

Teaching places great emphasis on interaction among students and between students and teachers, and independent studies. On the day of our visit, a group of students were studying artificial rearing of Coccinellid (*Coccinella septipunctata*) and examining beetles collected from a cotton field.

Peking University offers night courses, short courses, and correspondence courses. It organizes courses in factories and on farms that are comparable with training courses on specialized subjects.

Northwest College of Agriculture The Northwest College of Agriculture was built in 1934 as a small institution with three laboratories and has expanded to the present 48 laboratories. The college has 1,500 staff members and 2,000 students in seven departments: agricultural economics, agricultural engineering, agronomy, horticulture, hydrology, plant protection, and veterinary medicine. The relatively high ratio of staff is due to the fact that the college is also a research organization. The faculty also teaches in factories and on farms.

Chungshan University Chungshan University, established in 1924 as Kwangtung University, was renamed in 1929. It has 11 departments:

biology, chemistry, Chinese, economics, foreign languages, geography, history, mathematics, metallurgy, philosophy, and physics.

There were 4,700 students before the Cultural Revolution. Now there are about 1,000 teachers, 1,000 staff members, and 2,600 students. The university expects 5,000 by 1980. It also offers short courses that were attended by 45,000 students in January through June of 1975. The biology department has subdepartments of entomology, genetics, physiology, and zoology. The department has 200 teachers and staff and 350 students. The teachers regularly spend 4 months each year in the countryside with students. This experience is to relate theories to production. In the spring of 1975, a class of 40 third-year entomology students went to the Big Sand Commune near Kwangchow. They were divided into 10 groups, 4 general (teaching material collection, medical insects, *Bacillus thuringiensis,* and insect viruses) and 6 crop-oriented. Each group stayed with a production brigade and worked closely with its farmers.

Philosophy of Pest Control Training

This is an account of plant protection training by a third-year student of entomology at the University.

"I came to the commune (Tahsia [Big Sand] in Ssu-hui County) in April 1975 to participate in rice integrated control. My re-education was under the leadership of the party committee and I have made some progress ideologically and educationally. The students were divided into 10 small groups for research purposes: insect viruses, biological control, *B.t.* (*Bacillus thuringiensis*), mathematical evaluation, and important pests (six groups). Each group stayed in a production brigade. My group studied the rice thrips, *Thrips oryzae* Williams. We met with farmer technicians every week for 1 to 2 days for survey work with production teams, spraying, and other production labor in order to relate our research to field practice. In addition, we had 2 days for insect rearing and identification in the laboratory, and 2 days to study theory. According to Mao's teaching, "the bourgeoisie will be re-educated by the peasant," so every week we joined with the poor and low-middle peasants for one or two nights of 1 to 2 hours of political study of Marxism and Leninism. Our teachers also lectured on pest conditions in the field, on pest forecasting, and on plant pathology.

"University study should serve proletarian politics and unite worker, peasant, soldier, and student. In these months we studied proletarian dictatorship theories with the farmers and enhanced our ideological consciousness and understanding of Mao's theories. Many students were criticized for looking down on farm work and experienced a determination to go to the country after graduation and to serve the peasants all their lives. This study diminished the differences between workers and peasants, city and country people, and mental and manual work.

"New students need to develop correct ideas when they go to the university, so research must be combined with practice to develop agriculture. Every one of the small research groups took up a practical problem. For example, the *B.t.* group studied how to produce *B.t.* in indigenous locations, economically and quickly. The *Trichogramma* group studied the preservation of the vitality of the stock during rearing. The thrips group studied the life history and its transfer to the fields, and looked for effective insecticides.

"According to Mao's teaching -- launch a mass movement. After I came to the commune I had a lot of meetings with plant protection teams to analyze insect behavior, scientific plans, control measures, survey of pest species, characteristics of thrips, egg-laying, etc. Thrips are hard to control with insecticides because of difficulty in proper timing of applications, and many applications do not resolve the problem. Analysis of weak points of the thrips suggested insecticide applications during the nymphal stage, and experiments with one or two applications of insecticide (dimethoàte) gave control. We were helped by the peasants in indoor and outdoor experiments, and farmers sent all kinds of insecticides for evaluation. Thrips are tiny and hard to observe, but with the help of the peasants during the 3 months, we can identify species and understand life histories.

"One cannot find out anything by experimentation divorced from practice. The purpose of our work is the dominant question. In older universities where the revisionist educational line was carried out, students studied strange problems divorced from practice, read lots of books, and shut themselves in their rooms. We should change this completely, especially when we go to the country to learn the revolutionary spirit from the peasants. Thus, scientific research is changed from very small problems to important practical problems. The students spread the knowledge of integrated control throughout the communes and give lectures about insect control to the plant protection teams. In this way we promote the continuing development of plant protection. When pest conditions are observed, we advise the plant protection teams. When we meet new kinds of pests, we discuss them and will look at books to find new ways to control them. In this way we learn through practice, and teachers give lectures in the field and plant protectors give their experiences. Thus, we can master plant diseases and insect pests of rice in 2 months. We feel that courses of study must be integrated with practice to learn firmly and remember firmly. Our work is preliminary and can be improved."

Postcollege Education

Presently, there is no formal graduate school. Promising and academically oriented graduates are generally retained in the university or institute as research workers. They work with more senior members, gradually developing their own expertise.

These students receive no degree to mark the completion of a phase in their academic development, but simply become, after a period of time, regular researchers or teachers. Since the Cultural Revolution, ranks have been abandoned.

As will be mentioned elsewhere in this report, Chinese entomologists

keep up with foreign entomology literature, from which they derive
research data for application in local settings. Thus, foreign language
capability, especially in English, becomes crucial. A concern was
sensed that the younger scientists need to increase their English capability, especially if they are to interact with foreign scientists.

In general, teachers receive a salary twice that of peasants, plus
free campus housing. Otherwise they enjoy no greater social benefits
than other sectors of the society. There is no compulsory retirement
age. Mental workers (e.g., teachers) can retire at 60. Since the retirement pension is the same as the regular wage, a teacher beyond 60
often continues to work at a suitable pace.

Summary Observations

College curriculums have been trimmed to the most relevant disciplin-
ary courses. Much of what is considered liberal education in the United
States has been sacrificed. At the same time, there is a heavy emphasis on political training. With the emphasis on practical courses,
college education is hardly distinguishable from what would be considered vocational training in the United States.

College programs have been meshed with production. In this system,
internship comes during, rather than after, the formal education programs.

Some college programs in China are consistent with the concept of
university without walls and outreach programs in the United States.

Postcollege education is much less structured than in the United
States system and more liberalized than in the English system.

Many entomologists trained under this system are now productive in
research and teaching. How this system compares with more structured
and regimented systems elsewhere will not be known perhaps for another
academic generation. Since the Chinese system is considered as an
academic experiment, the results should be of interest to scientists
everywhere.

F.   INSECT TAXONOMY AND COLLECTIONS

Taxonomy

Insect taxonomy is an active field in China. In spite of the emphasis
on practical work, taxonomists are free to pursue their special interests. Drs. Chen Shih-hsiang and Chu Hung-fu, Director and Deputy Director, respectively, of the Institute of Zoology, Academia Sinica, Peking,
are active workers in spite of their administrative responsibilities
in a 500-member institution. At the same time, many young China-trained
workers are making significant contributions. The taxa dealt with in 9
recent issues of *Acta Entomologica Sinica* (1973-1975) and the institutions involved are as follows:

Taxa	No. of Papers	Institutions
Acarina	5	Shangung Medical College; Department of Biology Kiangsi University; Shanghai Medical College Museum of Natural History, Shanghai Institute of Zoology, Peking
Protura	1	Institute of Zoology, Peking
Orthoptera	2	Department of Biology, Shensi Normal University Prairie Work Team, Chinghai Institute of Biology, Chinghai
Zoraptera	1	Institute of Zoology, Peking
Isoptera	2	Institute of Zoology, Peking
Neuroptera	1	Peking Agricultural University
Anoplura	2	Kweiyang Medical College
Hemiptera	3	Department of Biology, Nankai University, Peking
Homoptera	5	Institute of Zoology, Peking Szechuan Academy of Agriculture Agricultural College Anhwei Agricultural College
Coleoptera	8	Institute of Zoology, Peking Kwangtung Institute of Entomology
Lepidoptera	2	Northwestern College of Agriculture Institute of Zoology, Peking
Diptera	11	Institute of Zoology, Peking Kwangtung Health Station Sanitation and Disease Control Station, Kweichow Academy of Medical Science, Peking Institute of Parasitic Diseases, Shanghai
Siphonaptera	11	Epidemic Prevention Station, Kangting Szechuan Health Center Institute of Epidemic Disease, Yunnan Kweiyang Medical College Health and Anti-epidemic Station, Kweiyang

Taxa	No. of Papers	Institutions
		Institute for Control of Local Diseases, Chinghai
		Fukien Providence Anti-epidemic and Health Station
Hymenoptera	2	University of Agriculture and Forestry, Fukien
		Institute of Zoology, Peking
		College of Agriculture and Forestry, Kwangtung
Fossils, insects	2	Institute of Zoology, Peking North China Geological Institute

This summary shows the variety of taxa covered and the wide distribution of institutions making contributions. The health emphasis is reflected in the large number of papers under Diptera (Culicidae), Siphonaptera, and Acarina. Of the eight papers under Coleoptera, five were contributed by Dr. Chen.

Multiple authorship is common. The authors of a given paper may be from different provinces, reflecting interprovincial cooperation. All together, 71 authors are involved, representing a healthy pool of taxonomy expertise.

Dr. Chu indicated that specimens are being received by the Institute of Zoology from all parts of China. The institute also has expedition teams. Three teams have been sent to Tibet. Travel is apparently not difficult for scientific purposes. Dr. Yang of the Shanghai Institute of Entomology went to Yunnan Province to collect scale insects of pine for the sole purpose of clarifying some taxonomic uncertainty.

Insect Collections

The insect collections of four institutions were visited. The collections were generally good. In the South, high humidity is a problem, especially since the cabinets are wooden.

- Institute of Zoology, Peking There are an estimated 1,500,000 to 2,000,000 specimens. Groups especially well represented are Chrysomelidae, Curculionidae, Scolytidae, Isoptera, and Macrolepidoptera.
- North China College of Agriculture, Sian There are an estimated 15,000 species, with 5,000 local forms.
- Shanghai Institute of Entomology, Shanghai There are about 200,000 specimens, mainly from eastern China. However, Coccidae, Protura, and Isoptera are from other regions. Many of the specimens bore labels indicating that they were collected by the American entomologist J. L. Gressitt.
- Chungshan University, Kwangchow There are about 200,000 specimens. Groups especially well represented are Acrididae, Cerambycidae, Pentatomidae, Hydrophilidae, and Dytiscidae. There are 20,000 specimens of the last 2 families.

G. ENTOMOLOGICAL LITERATURE AND LIBRARIES

Library Resources

Four libraries were visited: Institute of Zoology, Academia Sinica, Peking; Peking University, Peking; Shanghai Institute of Entomology, Shanghai; and Zoology Department of Chungshan University, Kwangchow.

Institute of Zoology This institute has a very large holding of ento-mological and other zoological literature.
    Chinese journals noted were as follows:

● National scope. The following is a fairly complete list of the journals related to zoology published as national publications in China:

>   Chinese Science
>   Chinese Agricultural Science
>   Acta Entomologica Sinica
>   Acta Botanica Sinica
>   Acta Genetica Sinica
>   Acta Zoologica Sinica
>   Applied Microbiology
>   Soil Science
>   Fisheries Technology
>   Tropical Crops
>   Science and Technology
>   Agriculture and Forestry
>   Agricultural Science and Technology
>   Scientific Literature
>   Botanical Gazette
>   Forestry
>   Environmental Protection
>   Scientific Experimentation
>   Catalogue of Foreign Scientific Literature
>   Agricultural Chemical Industry
>   Agricultural Chemicals

● Provincial scope

>   Kwangtung Agricultural Sciences
>   Chekiang Agricultural Sciences
>   Hupei Agricultural Sciences

● University scope

>   Peking Normal College Bulletin
>   Langchow University Bulletin

● Chinese translations

>   Science (United States)
>   Nature (Great Britain)

Foreign language journals

● There are 12 Japanese journals. They include:

  Entomology
  Entomological Review
  Sericulture
  Agricultural Chemistry

● There are 111 zoological journals in other languages on the shelves (not including Russian, which were not checked). Of these, 43 deal with entomology and pest control:

  Acta Entomologica Bohemoslovaca
  Acta Entomologica Jugoslavica
  Annales de la Société Entomologique de Québec
  Annals of the Entomological Society of America
  Bulletin of the Entomological Society of America
  Canadian Entomologist
  Deutsche Entomologische Zeitschrift
  Entomologie Experimentalis et Applicata
  Entomological News
  Entomologische Blätter
  Entomologische Zeitschrift
  Entomologists Record
  Entomologica Americana
  Entomologische Meddelesen
  Entomophaga
  FAO Plant Protection Bulletin
  Folia Entomologia Hungarica
  Insect Biochemistry
  Insect Morphology and Embryology
  International Pest Control
  Journal of Economic Entomology
  Journal of Entomology (Royal Entomological Society of London)
  Journal of Stored Produce Research
  Journal of the Entomological Society of South Africa
  Journal of the Kansas Entomological Society
  Journal of the Lepidoptera Society
  Metteilungen der Entomologische Gesellschaft
  Mosquito News
  New Zealand Entomologist
  Notulae Entomologicae
  Nouvelle Revue d'Entomologie
  Pan-Pacific Entomologist
  Pacific Insects
  Oriental Insects
  PANS
  Pest Control

Pesticide Science
Proceedings of the Entomological Society of Washington
Tijdschrift voor Entomologie
Transactions of the American Entomological Society
Transactions of the Royal Entomological Society of London
Transactions of the Shikoku Entomological Society
Zeitschrift fur angewandte Entomologie

We do not know if each of these journals is complete. Many are obtained on an exchange basis. Some multilith copies were noted, e.g., *Scientific American* in one library, *Annals of the Entomological Society of America* in another, and *Journal of Economic Entomology* in another. American journals noted were only 1 or 2 issues behind.

The library has about 70,000 books. The holdings of Chinese-language entomology books were sampled. In a group of 145 titles covering a variety of areas, the distribution of publishing dates was as follows:

Pre-1949	5	1957	15 (1)	1967	1 (1)
1949	2	1958	15 (1)	1968	0
1950	5	1959	17	1969	0
1951	8	1960	7	1970	0
1952	1	1961	6 (1)	1971	0 (5)
1953	2	1962	5 (1)	1972	2 (16)
1954	9	1963	12	1973	0 (8)
1955	6	1964	10 (2)	1974	1 (12)
1956	9 (1)	1965	6 (2)	1975	0 (9)
		1966	2 (5)		

A number of entomology books were purchased in local bookstores, and will be considered later. Their numbers are given in parentheses in the above table. Examining books in the libraries and those available in the bookstores one finds three lean periods. It is interesting to relate these to social or political developments: 1952-1953 was during the early constriction of private enterprise, 1960-1962 immediately followed the period of economic difficulty, and 1967-1970 was the period of Cultural Revolution. It is gratifying to note that issuance of entomological publications has since resumed a vigorous pace, especially in the practical areas. Prior to 1959, many Russian books were translated, but none since then were noted.

Library of Peking University The library of Peking University is the newest and most spacious in China. Unfortunately, time did not permit visiting its entomology holdings.

Libraries of Shanghai Institute of Entomology and Chungshan University These libraries are not as extensive as the library of the Institute of Zoology in Peking. But they do have additional publications, including some of a provincial nature, such as issues from Futan University, Anhwei University, and Chungshan University.

The Leading Entomological Journal in China

Acta Entomologica Sinica is the major national entomological journal in China. Another is *Entomological Knowledge,* containing popular and practical articles.

The Acta is edited by the Chinese Entomological Society and published by the Scientific Press. Domestic sale is handled by local post offices, and foreign sale by the Chinese International Book Store. This journal was started in 1950 with annual volumes being printed, except during 1951, and during 1967-1972. Issues in Volume 10 were published in 1960-1961. The lapse from 1967-1972 coincided with the period of restructuring of educational and scientific systems on a national scale. Publication was resumed in 1973 with only 2 numbers appearing that year. In 1974, the 4 regular numbers were published. In 1975, 3 numbers had been issued at the time of our visit. We were told that number 4 was expected to be on schedule. English translations of this journal were being published by Plenum Publishing Corporation, New York, under the editorship of Dr. Shen Chin Chang, U.S. Department of Agriculture, Beltsville, Maryland.

The 9 issues that had appeared since the resumption of publishing were made available to the delegation. Their contents are summarized as follows:

	1973	1974	1975
Volume	XVI	XVII	XVIII
Number of issues	2	4	3
Chairman Mao's statements	0	4	3
Political articles	0	5	4
Numbers of papers in different areas			
Taxonomy	10	37	16
Morphology	1	2	1
Physiology	4	4	2
Bionomics	5	10	6
Applied ecology (including forecasting)	0	3	2
General economic entomology (including integrated control)	0	3	7
Biological control	3	6	8
Chemical control	2	3	5
Other economic aspects	0	2	1
Total	25	70	48
Number of papers with team authorship	3	10	19
Number of papers with English abstracts	18	52	23
Number of papers with Russian abstracts		1	
Number of papers with German abstracts			1

Several aspects of this summary may be worth mentioning.

● Starting with Volume XVII, each number is prefaced by statements by Chairman Mao Tsetung. Some relate to the importance of natural sciences and experimentation but most are strictly political.

● Also starting with Volume XVII, there are 1 or 2 articles in each number analyzing contemporary political issues. These articles are prepared by special political caucuses of production or research organizations. The inclusion of Chairman Mao's statements and political articles in scientific journals is, in China, consistent with the general concept that all the people, including scientists, must be conscious of politics.

● There is an increasing number of papers dealing with applied entomology with authorship by a research team or unit rather than by individuals. In some cases, two research units are involved, such as a research institute and a production unit (commune or brigade). The significance of this point is discussed elsewhere.

● Papers in taxonomy outnumber those in other fields. This is understandable in view of the vast area and vast fauna of the country and the relatively recent emphasis on pest control technology.

● A great many articles published later deal with material (specimens and experimental data) collected in the 1950's and 1960's. This may be a reflection of the lack of outlets for publication, especially from 1967 to 1972. It is gratifying to see that this material is now being published for posterity.

● Of the 143 papers in the nine issues, 95 carried abstracts in foreign languages. Of these, one was in Russian, one was in German, and the others were in English. Some of these, especially taxonomic ones, are quite detailed.

Editing of this journal is handled by an editorial committee, currently headed by Dr. Chu Hung-fu of the Institute of Zoology, Academia Sinica, in Peking. Manuscripts submitted are reviewed by peers. Since there have been more and longer manuscripts in taxonomy than in experimental areas, a quota system has been instituted. As a result, the waiting list for taxonomic papers is longer than the others. A separate publication for systematic entomology has been contemplated.

To get a glimpse of how up-to-date Chinese entomologists are in English entomological literature, 14 papers in Volume XVIII of Acta (1975) citing both Chinese and English references, are compared below.

Most recent Chinese reference cited in the paper	Most recent English reference cited in the paper	The publication in English
1974	1972	J. Invert. Path.
1964	1971	J. Econ. Entomol.
1965	1971	a book
1957	1961	a book
1961	1967	J. Med. Entomol.

Most recent Chinese reference cited in the paper	Most recent English reference cited in the paper	The publication in English
1973	1971	IAEA, Vienna
1973	1971	J. Econ. Entomol.
1964	1971	Sci. Pest Control
1957	1967	System. Zool.
1974	1971	J. Econ. Entomol.
1974	1973	Pesticide Sci.
1974	1972	Agri. Biol. Chem.
1973	1969	J. Gen. Virol.
1965	1962	Proc. Ent. Soc. Wash.

This comparison does not tell us anything about the extent of the Chinese entomologists' use of the literature in English, but it does suggest that their use of English references is comparable in recentness with their use of Chinese references. It also indicates that a variety of English-language journals are cited.

Currently Available Entomology Books

An effort was made to purchase entomology books in China. They are published by many publishers, but their sale is handled by one agency, the Hsing-hua (New China) Bookstore represented in all the cities visited. Most of the books available are paperback editions and deal with economic aspects of entomology. A series of 8 handbooks on the control of insects of different crops is available, as is a composite edition. A series of 6 pictorial guides is available in the same way. A person may buy any number of components or the composite volume.

A total of 67 books were collected: 8 presented by Chinese colleagues, 59 were purchased. The 67 included all entomology books available in the New China Bookstores in Peking, Shanghai, and Kwangchow at that time. Following is a list of the subjects treated and the number of books devoted to each.

General agriculture	4
General entomology	8
Pictorial guides	5
Handbooks	5
General insect control	3
Agricultural and forest pest control	
Rice insects	5
Grain crop insects	5
Cotton and peanuts insects	4
Soil insects	3
Fruit tree insects	2
Other tree insects	4
Stored products and structural insects	2
Pesticides	5
Beneficial insects	10
Plant diseases	2

There were many books on general agriculture. Only 4 were purchased there because of special interest, one of which was *Handbook of Agricultural Techniques for State Farms,* published in May 1975. The information in this 632-page book was compiled for the use of cadres, workers, and youths assigned to state farms. This comprehensive book may be indicative of the importance placed by the national government on the state farms in the Chinese agricultural system. Another on general agriculture was *Experimentation in Agricultural Sciences by the Masses.* This 87-page book gives 9 examples of experimentation initiated by the masses (in general, peasants on the farms) that resulted in increased production in widely scattered locations. Participation by farmers in agricultural research and experimentation is vigorously promoted.

The 8 books on general entomology include 4 of a series of 10 volumes entitled *Chinese Economic Insects* and 1 of a series of 3 volumes entitled *Insect Taxonomy.* It was surprising to find few such general books in the stores. In fact, the 4 volumes on economic insects were presented to us by the Institute of Zoology when officials there learned that the series was not available in the stores.

All the other publications are of a practical nature. Many have excellent color illustrations of insects and their damage.

It is of interest that under "beneficial insects," honey bees, silkworms, lac insects, wax insects, and the like are designated as resource insects, as are parasites and predators.

Book prices are very low. For example, the *Handbook for Agricultural Techniques for State Farms* costs 70 cents, and the 158-page *Pictorial Guide to Moths* (including 48 color plates) costs $2.60.

Virtually all these books except those on general entomology are prepared by committees rather than by individuals.

The general entomology books are published by the Scientific Press and the practical books by the Agricultural Press, the People's Press, and others. The Scientific Press and Agricultural Press are national and are located in Peking, while many provinces have branches of the People's Press. There are seven provincial, two municipal, and one regional People's Press represented in the books we collected: Shanghai and Tientsin Municipalities; Hunan, Hupei, Kwangsi, Kwangtung, Shantung, Shensi, and Szechuan Provinces; and Inner Mongolia. Other publishers include Shanghai Science and Technology Press, the Petroleum Industry Press, Fuel Industry Press, and Technical Standards Press.

2

INSECT PROBLEMS BY CROPS

OR PRODUCTS AFFECTED

A. RICE

General

Rice remains China's most important crop, essential for providing much
of the basic nutrition for its more than 900 million people. In 1975,
rice was sown on an estimated 35 million hectares (52,823,315 mu). This
is expected to produce an estimated 109,200,000 tonnes of grain. We
observed rice production in every area visited in China.

In northeastern China, rice is limited and is far less important than
soybeans, wheat, corn, sorghum, millet, and vegetables. In Hopei Province, rice joins important fields of wheat, barley, corn, millet, soybeans, and vegetables. In Shensi Province, rice is grown primarily in
the more southern areas (the Han Shui River Valley). Very little is
grown in the Wei Valley, in which Sian is located. Rice observed growing in southern Shensi is of the *indica* type, whereas that around Sian
is mostly *japonica*. Rice is rotated with corn. Often a triple crop
per year is obtained by using a rotation of corn-rice-vegetable (or
rape).

In the Shanghai area, rice dominates land use, with two crops produced annually, usually followed by a third crop of vegetables or rape.
In this area, the early rice is mostly of *indica* varieties, and the
late rice is of *japonica* types. In Hunan Province, rice predominates;
huge land areas are covered by this crop. Two crops per year is standard, with a third crop of rape or vegetable produced in the winter
months on most of the land. Most of the rice is of *indica* varieties,
developed for their short straw, early maturity, and high yield.

We were told that the *indica* types contain greater resistance to
certain plant diseases and are to some degree more tolerant to insect
pests because of their fast growth and rapid maturity.

Rice disease and insect problems are less severe in northern and
central China, more severe in the South. In the southern Yangtze basin
and in the subtropical and tropical South, insect pests are abundant,
constant, and varied. Cultural, biological, and chemical control methods are used to ensure high productivity of the fields.

Little mechanization was noted in rice production. We saw a mechanized rice transplanter in Shanghai and one at work on the grounds of the
Shanghai Academy of Agricultural Sciences, but none in the countryside.
In the northern central rice-growing areas, oxen are the primary animals
used in rice paddy cultivation. In the southern Yangtze basin and especially in the subtropical and tropical South, the water buffalo remains the primary animal used in rice production.

Weed control and cultivation are primarily by hand. Few chemicals
are applied for weed control in the rice paddies. Water buffalo are
used extensively to graze the small levees in the rice fields and the
drainage ditches and canals.

All fields are exceptionally clean of weeds and in good condition.
Most of the rice seen in the South was the late crop of rice, which was
transplanted in June or July. We saw fields in the Kwangchow area being
transplanted during our stay there, August 25 - 28, 1975. These were
mainly low-lying fields and pond bottoms where water was stored earlier.

There appears to be adequate water for good rice production in all
the areas visited. One of the "miracles" of Chinese agriculture has
been the development of sophisticated irrigation systems, land reclamation, and subsequent management of land and water resources. These
practices and development of rapid-maturing, high-yielding, short-
stalked rice varieties, ideally suited for double- and triple-cropping
systems, have provided reliability for sustained high yields of rice
in the past several years. We were told that the objective for the
southern rice area of the country was seven crops (five rice, two other)
within a 2-year period.

Fertilization of rice has greatly increased in the past few years.
We did not observe any fields with severe nutrient deficiencies. The
roads we traveled were lined with carts, bicycles, and wheelbarrows,
etc., carrying "night soil" to the fields, and commercial fertilizers
from numerous small fertilizer plants were evident along the road. Fertilizer was applied in the paddies in liquid form (hauled to fields in
approximate 35-liter jars) and in dry granular form. Lime was spread
by hand from the windward side of the paddies. The wind would float the
lime dust across the paddy in a more or less even pattern.

"Mudballing," incorporating fertilizers in mudballs and placing the
balls at the roots of the rice plants, is a widely used technique in
southern China. Although very time-consuming, this is a very effective
way to get fertilizer to the root area with minimum nutrient loss.

We were told that China invested over 2 billion yuan to build fertilizer plants in the countryside from 1974 to 1976. It was hoped that
this step would end the need to import expensive fertilizers. One of
the primary fertilizers currently made is ammonium bicarbonate, which
supplies much of the needed nitrogen source. We passed several of these
roadside factories where coal dust was being hauled in great quantities
from railroads, primarily by human-drawn carts. Coal and water are the
only raw materials required to produce the ammonium bicarbonate. Because of the simplicity of production, the factories can easily be
built in the countryside close to the crops.

Insect Pests of Rice

At least 89 species of insects attack rice. The more important are
the rice paddy borer, *Tryporyza incertulas* Walker; rice stem borer,
Chilo suppressalis Walker; purplish stem borer, *Sesamia inferens* Walker;
brown-backed planthopper, *Nilaparvata lugens* Stäl; white-backed planthopper, *Sogata furcifera* Horvath; grass leafroller, *Cnaphalocrocis
medinalis* Guenee; and rice case butterfly, *Parnara* spp.

The most severe pest groups are rice stem borers, rice planthoppers,
rice leafhoppers, and grass leafrollers. There are several species of
each, which makes control difficult.

In Hunan Province, at the Institute of Plant Protection, we were
told that several insects that had been major pests in the past have
been basically controlled. These included the rice grasshopper, black
grasshopper, rice bug, rice leafbeetle, rice rootworm, and army weevil.

Changes in cultivation and management of rice in recent years, involving greater use of commercial fertilizer, better water management,
new varieties, and intensive double and triple cropping, have greatly
increased the occurrence and severity of rice pests. Examples given
were the increases in occurrence and damage of pests such as the rice
leafhopper and the rice planthopper.

Increased abundance of rice leafhoppers has increased the occurrence
of plant viruses transmitted by these insects. Insects such as the
grass leafroller and the grass thrips, once minor or secondary problems,
have become primary pests. Failure to control these pests has decreased
rice yields.

Biology of Rice Stem Borers

In the major rice-growing provinces of Hunan and Kwangtung in the
South, there are three species of rice stem borers: the rice paddy
borer, the rice stem borer, and the purplish stem borer. Prior to 1957,
these species were not severe pests.

As acreage increased with reclamation of new land and the beginning
of double-cropping, the rice paddy borer and the purplish stem borer
greatly increased and spread into areas where they had not previously
been pests. But the rice stem borer has decreased in importance.

The rice stem borer has three generations a year and the harvest
season of early rice occurs in middle and late July during the occurrence of the third to fourth instars of the second generation. The
larvae of the rice stem borer are killed during harvesting and subsequent handling of the straw. As a result, damage to late rice by this
pest is usually very light.

The rice paddy borer in Hunan and Kwangtung provinces and other
southern areas of China has four generations a year. It is monophagous,
feeding only on rice. It causes the greatest damage in the third and
fourth generations. The third generation occurs in middle and late
August during the rice tillering period. The damage causes "dead heart"
and subsequent severe reduction in yield. The fourth generation occurs

in middle and late September during time of fruiting and causes "dead head." Without control, damage can reduce yield by 80%.

We were told that before Liberation, damage by rice stem borers averaged more than 40% and that damage has been reduced to less than 1% since the Cultural Revolution. Although we could not verify this, it was evident that in the paddy fields visited, damage by rice stem borers was very light.

Biology of the Green Rice Leafhopper, *Nephotettix cincticeps*, and Brown-backed Planthopper, *Nilaparvata lugens*.

The leafhoppers and planthoppers of rice produce 6 or 7 generations a year in southern China. With the advent of double-cropping, abundant food has been provided over a larger part of the year. The abundant food, high production capacity, and migratory habits of these insects have caused them to become serious pests on rice. They inflict severe damage in the rice seedling beds, then migrate to the early transplanted rice crop. After harvesting of the early rice crop, populations migrate to the late crop after the transplanting.

The green rice leafhopper, *N. cincticeps*, is the vector of several serious rice virus diseases.

Chemical Control

The proper time to apply chemical control is determined through a pest forecasting system set up at four levels: the county, the commune, the production brigade, and the production team. (See "Extension Programs" in Chapter 1.) There appear to be trained plant-protection specialists at each level.

There is also a district forecasting laboratory (usually at an institute), which studies the occurrence, population dynamics, and economic damage levels of the pests and the potential impact of natural enemy species. The county forecasting station determines time of occurrence of pests through the accumulation and assimilation of data obtained from the commune and brigade forecasting teams.

The county forecasting station advised the commune and brigades when to carry out control operations. Commune and production brigades make final decisions on whether to make applications. Decisions are based on local conditions, presence of pests and natural enemies, weather, and so on. Application of insecticide is by specialized plant-protection teams at the production brigade level. In cases of heavy pest outbreak, the masses in the commune may be mobilized to help in implementing controls.

Pests of Rice Seedling Beds

Leafhoppers, rice paddy borers, and other minor pests are controlled by two sprays--DDT or dimethoate emulsion--in early rice beds. Fenitrothion is also used. In late rice beds, three or four sprays of these materials may be needed. Trichlorfon and fenitrothion are applied in late rice-seedling beds.

Pests of Rice Paddies

The main pests to control in middle to late May are rice leafhoppers, black leafrollers, and first-generation rice stem borers. Secondary pests during this time are rice thrips and rice skippers.

From late June to early July, the main pests are planthoppers, second-generation leafrollers, rice thrips, and certain secondary pests. Insecticides used for these pests in paddies are mixtures of methyl parathion and BHC. In middle and late July, the primary pests (in paddies) are third-generation leafrollers, leafhoppers, and certain secondary pests. In early and mid-August, the main pests are third-generation rice paddy borers, rice skippers, and rice planthoppers. In late September, as the rice matures, the main pests are fourth-generation rice paddy borers and *Mythimna separata*. Insecticides used in the late rice fields are methyl parathion (plus BHC), trichlorfon, and fenitrothion.

We were told that within the production brigades are indigenous insecticide factories. These factories produce such insecticide materials as tobacco and lime powder and lime plus garlic. These products are used against eggs of the rice paddy stem borer. Also mentioned was an insecticide, similar to rotenone, derived from a plant.

Cultural Control

Good cultural practice is the foundation of insect control in China. The following are some of the more important cultural practices in the major rice-growing areas in the South:

- Weeding Planthoppers and leafhoppers are destroyed in overwintering sites, which helps protect the late rice from the insects and virus diseases.
- Trapping A small section of a field is planted early to trap insects, which are then sprayed before the rest of the field is planted to rice.
- Planting early maturing varieties (*indica* types) Planting is done before June 26 to avoid damage by second-generation rice paddy borers. Rice is planted early so it will mature before September 20 to avoid damage by fourth-generation rice paddy borers.
- Cultivation and manual control Fields and adjacent areas are kept well cultivated, and insects and insect eggs are removed manually and destroyed when observed.
- Flooding At Tahsia (Big Sand) Commune, in Kwangtung Province, flooding is used to help control rice paddy borers. Rice fields are immersed in several inches of water prior to March 6 each year to drown larvae in rice stems--a very effective technique in that area.

Physical Control--Blacklight Traps

In nearly all the rice-growing areas of China we visited, light traps were extensively used as a monitoring or survey device for the pest forecasting network. In Hunan Province, we were told that blacklights are also effective as a direct suppression in helping to control such rice pests as the rice paddy borer, the rice leafroller, planthoppers, leafhoppers, and lawn armyworms.

It was indicated that 85% of the moths of the rice stem borer and rice paddy borer were being attracted to the traps before mating and oviposition. Hunan Province had made a major commitment to the concept of controlling rice insects by use of blacklight traps. All the rice acreage had traps, with an average of about one trap per 8 or 9 acres.

We were told that over 1 million traps had been installed and were operational in this province, which is clearly the world's largest light-trapped area. We have some skepticism as to how effective the traps are. No data were shown to prove their worth. Our doubts are based primarily on a long history of U.S. research showing that light traps are ineffective in economically controlling certain species of Lepidoptera, such as the cotton bollworm, tobacco budworm, tobacco horn-worm, and pink bollworm. In adjoining Kwangtung Province, light traps are not being used in rice as a direct pest-control effort, although some are used for monitoring. A Chinese entomologist in this province indicated that their data show that sometimes more beneficial insects were trapped than pest species. They were continuing to research light traps but seemed to have strong doubts concerning their overall effec-tiveness as a direct control technique. They also said there were eco-nomic problems because of costs of running electric lines through the rice fields, investment in traps, and subsequent maintenance.

Biological Control

The use of biological control agents for insect control varies from none to considerable use of *Trichogramma* and insect pathogens such as *B.t.* In Kwangtung Province--at the Entomology Research Institute and at Chungshan University--some basic studies were being conducted on biology, ecology, population dynamics, and efficiency of natural para-sites and predators, including spiders. (See also "Biological Control" in Chapter 4.)

Integrated Control

The most sophisticated integrated pest-management system that we saw was located at Tahsia (Big Sand) Commune about 150 km from Canton. The system consists of the following:

● Insect forecasting network.
● Flooding of rice fields prior to March 6 to drown larvae of the rice paddy borer.

- Blacklight traps, one trap per 2.31 hectares (35 mu).
- Use of ducks in rice fields for consumption of leafhoppers, plant-hoppers, and rice leafrollers. The average number of ducks per mu is three. However, the ducks are herded through the rice paddies in large numbers.
- Use of frogs in the rice paddies for insect suppression.
- Use of a rice variety resistant to a fungus leaf disease (*Rhizoctonia*).
- Release of *Trichogramma* wasps for control of the rice leafroller.
- Utilization of *Bacillus thuringiensis* and reduced amounts of the insecticide dimethoate for control of rice paddy borer and other Lepidoptera, such as the rice leafroller.

Commune insecticide costs per acre were reduced by this integrated program from more than 3 yuan to 1.03 yuan ($1.50 to $0.52).

Integrated control for most of the rice-producing areas of China seems still to consist primarily of pest forecasting plus good cultural control and use of chemicals. Other techniques, such as use of parasites and pathogens, may be used in the future on larger acreages as techniques for their production are worked out and the methods become popularized. Yet, without question, integrated control is being used more widely on rice in China than on any other major crop in the world. This is possible because of the national emphasis on agriculture and its organizational structure, especially at the commune level and the abundant labor available to implement integrated control practices. The intensive Chinese agriculture, which has been developed over centuries, has evolved many cultural practices that, when properly managed, can help suppress insect pests.

Concluding Remarks

In the foreseeable future, rice will continue to be the main food crop of China. Success of the Chinese in increasing rice production through construction of sophisticated irrigation schemes and ambitious land-leveling and reclamation projects over the past 10 - 15 years is one of the country's greatest agricultural feats.

A second accomplishment has been the increased production of agricultural products through intense double- and triple-cropping. This has been made possible by development of fast, early-maturing short-stemmed rice varieties selected from native materials.

The tremendous achievements that have been made in rice production will make significant progress in the future more difficult but certainly not impossible. Much remains to be done in improving yield, uniformity, and quality of rice. This progress in the future will be possible only through a vigorous breeding program designed to decrease maturity time for double- and triple-cropping systems and emphasizing quality, yield, uniformity, and, especially, disease and insect resistance.

We did not visit many places where rice plant breeding was being conducted on a large scale. In discussions with entomologists and breeders at the stations visited, it was evident that China has not formed effective multidisciplinary teams of plant breeders, plant pathologists, and entomologists to speed the process of breeding for resistance to major diseases and insects. (See section on Status of Breeding for Resistance to Insect Pests in China). In fact, entomologists and plant breeders are often isolated from one another in entomology or plant breeding (genetic) institutes.

We were impressed with the stage of development of integrated control of pests in rice, especially the extent of use of cultural, biological, physical, and chemical methods. The success of the Chinese in practical implementation of biological and cultural pest-control methods can certainly be an inspiration to nations still struggling with the problems of implementation and especially to developing nations with similar problems.

B. COTTON

China is recognized as the third largest cotton-producing nation in the world with an average yearly output of about 10 million bales (a bale weighs 226.8 kilograms). Even so, cotton often is imported to meet the needs of the country. Thus, cotton is an important cash crop for the communal and state farms. The lint goes to meet the textile and industrial needs of the country and the seed is used for cooking oil and livestock feed.

Cotton is the clothing of choice in the summer in China. Men, women, and children were dressed in open-neck cotton shirts worn outside of cotton slacks. Everyone we saw was adequately clothed. We were surprised, however, to learn that clothing is rationed in China. The official purpose of rationing is to equalize distribution, not to cope with shortages. Even though clothing is rationed in China, recent U.S. reports (*Cotton Farming,* March 1976) indicated that more than 243 million m^2 of cotton fabric was exported from China to this country in 1974. The U.S. cotton trade also indicates that China is interested in purchasing raw cotton from the United States, provided that satisfactory agreements can be reached for exportation of finished cotton goods back to this country.

Distribution of the Crop

Unfortunately the delegation was not able to visit the major cotton-producing areas of China. This report, therefore, must draw on the literature as well as our impressions from one visit to provide information on cotton production in China. Efimov and Miftakhov (1954)[1] reported that more than 50% of all the cotton is grown in the regions north of the Yangtze River in the Great Chinese Plain, in the delta and middle course of this river, and in the provinces of Shansi and Shensi.

The above authors divided the cotton-producing regions of China into 4 zones according to the seriousness of the main pest at the time, the pink bollworm, *Pectinophora gossypiella* (Saunders). The zones are as follows:

First zone: This zone embraces the basin of the Yangtze River between latitude 28° and 34° north and longitude 105°-120° east. Included are the provinces of Chekiang, Hupeh, Szechwan, Anhwei, and Kiangsu and the northern parts of Shansi and Hunan. The temperature in this region seldom falls below 0° C and the climate is characterized by adequate, evenly distributed rainfall. Cotton is planted from mid-March to mid-April.

Second: This zone includes the mountain region of the Yunnan-Kweichow plain. It is not an important cotton-growing region in terms of total production.

Third: This is the region along the Hsi Chaing River, which flows through the provinces of Kwangtung and Kwangshi. It includes all the subtropical zone of China and the areas of the northern tropics. The summers are very hot (38° C), and winters may have days that are below freezing.

Fourth: This is the region of the Great Chinese Plain and surrounding mountains and includes the provinces of Hunan, Shantung, and Shensi and the whole of northern China.

A second report on the distribution of cotton in China was made by Kung (1975)[21]. He reported that the North China Plain and Shantung Hills (a region covering southeastern Hopei, Shantung, the northern part of Kiangsu and Anhwei, and central and eastern Hunan) produce about two-thirds of the country's crop. In this region, 68% of the land is cultivated and 10% of the cultivated land is irrigated. The region is characterized by the fertile soil of the Yellow River flood plain, an average annual rainfall of 58.42 cm, and an 8-month growing season. The average January and July temperatures are -3.8° C and 26.6° C, respectively. Single-cropping is the pattern. Maize, cotton, and sorghum are grown mainly in the warm months, and the land is left fallow in the winter. About 40% of the land is double-cropped. Wheat and barley are the principal crops grown in the cool months, and this region produces about one-half of China's total output of these grains. Kung reported that the Loess uplands of the western part of Hopei, central and southern Shansi, most of Shensi, the southern tip of Ningsia, and southeastern Kansu also produce some cotton, mainly in the valleys of the Fen River in Shansi and the Wei River in Shensi. Cotton does well in these river valleys, and the cotton production here accounts for about one-sixth of China's total. The land usually is planted in the spring and left fallow in the winter. Double-cropping is practiced on some of the land; cotton is followed by wheat or millet in the winter.

Cotton also is a major crop in central China along the 34th parallel between the Yellow and Yangtze rivers and drains the northeastern margin of the north plain of China. Cotton is the chief cash crop in this region, which is characterized by wet summers (an annual rainfall of 114.3-139.7 cm) and a growing season of 300-330 days. This allows for considerable double-cropping. Rice and cotton are grown in the summer, followed by wheat, barley, rape, beans, and green manure crops in the winter.

The rice and winter wheat area of the northern Yangtze basin and delta also is reported by Kung to be a cotton-producing area. The area is bordered by the Huai River in the north, the Yangtze in the south, the highlands west of the Han River, and the coastal area to the east. It also stretches south-southeastward, covering the lower Yangtze delta and the Hangchow bay. The whole basin includes the northern tip of Chekiang, central and southern Kiangsu, central Anhwei, central and eastern Hopei, and the Tungting Lake basin of Hunan. It is made up largely of flood plains and lakelands where crops are intensively cultivated. In July the temperature averages about 26.6° C and in January it may drop to 1.67° C. Average annual precipitation is 114.3 cm, ranging from 147.32 cm in the south to 76.2 cm in the Huai Valley. There is good summer rainfall. Because of this and the large numbers of people available, farming is highly intensive even by Chinese standards. Rice and cotton are the summer crops, followed by wheat, barley, and rape in the winter. Some soybeans, broad beans, field peas, buckwheat, corn, and sweet potatoes are also produced.

Kung (1975)[2] also reports cotton production in the southern Yangtze basin in Chekiang, Kiangsi, and Hunan provinces, in northern Fukien, and in southern Anhwei. This basin is watered by several streams that flow into the giant lakes of Tungting and Poyang and then into the Yangtze River. There is a dense network of irrigation canals and waterways in this region. The growing season is about 300 days. In July the average temperature is 29.4° C and in January 4.4° C. Annual rainfall ranges from 114.3 to 132 cm and is evenly distributed. High temperatures and rainfall make the summers oppressively humid. Rice is the principal summer crop, and Hunan is known as the rice bowl of China. Cotton is grown mainly in small irrigated plots that are planted to wheat or rape in the winter. The principal area of cotton production in this region is in the Poyang basin.

The Szechwan Basin also is reported by Kung to produce some cotton, but rice and wheat are the principal crops.

In the remainder of this section, observations made by the delegation concerning cotton production in China and the various reports given to us by Chinese scientists will be discussed.

In the Peking area, we saw many small fields of cotton, 1-7 hectares in size, in outlying acres north of the city. We were told that some cotton was grown in coastal areas as far north as 483 km from Peking. Peking, with a latitude of about 40° north, is characterized by a very warm summer and relatively cold winter. The cotton in this area was not interplanted with any other crops but was being grown in close row spacings, 35.5-45.7 cm, and under irrigation. The area did not appear to be a major cotton-producing area; the fields were scattered among rice, corn, sorghum, fruit, and vegetables.

In Sian, cotton was being grown in conjunction with corn, wheat, millet, and various vegetables. Corn appeared to be the main summer crop, but there was considerable cotton. The area is hot and arid and has an annual rainfall of about 51 cm. The cotton was being produced under irrigation and looked very good. The area is similar to the Trans-Pecos area of Texas. We estimated that the yield was 1 to 1-1/2 bales per acre. Cotton seedlings are transplanted to the fields from seedbeds in the middle of March. The first vegetative branches are pruned by hand. The plants are pruned twice more, with all nonfruiting branches being removed, and eventually are topped. Pruning and topping of cotton appear to be common practices in all the cotton-growing areas of China. We were told that the pruning was being done so that all the plant nutrients might be concentrated into the fruit-bearing branches.

A considerable amount of cotton is grown in the Shanghai area, where it is produced in conjunction with rice, wheat, rape, and vegetables. We visited one cotton commune where the average yield was 905 kilos of lint per hectare. All cotton is grown under irrigation and transplanted by hand in rows 35.5-40.6 cm apart. This is done in mid-March to mid-April, and harvest is from late September to early November. Rape for seed is grown in the winter. The following spring the cotton land may be replanted to cotton or rice. Cotton is often followed by winter wheat and then rice. All the land is multicropped.

During our visit to the Changsha area of Hunan Province, we saw small fields of irrigated cotton, but it is not a major crop there. Rice is the principal crop. All the cotton is irrigated and appeared to have a yield potential of 1-1/4 to 1-1/2 bales per acre.

Production Practices

In all the cotton-producing areas we visited, the crop was grown under supplemental irrigation. Water was applied as needed--two to four times a season. All crops appeared to be well fertilized. In Sian we were told that the following rates of fertilizer were applied per mu of cotton: 600 kilos of compost, 25 of superphosphate, 78 of urea, and 50 of waste cottonseed. Most of the cotton was hand-transplanted from seedbeds in high density plantings in rows 30.4-40.6 cm apart. The plants were hand-pruned an average of four times per season, three times to remove vegetative (nonfruiting) branches and once to top (removal of main stem terminal). All the varieties we saw in commercial production were American upland types, *Gossypium hirsutum,* that were similar to our delta types. However, in breeding plots and variety trials there were some long-staple *G. barbadense* types. We also saw in the breeding plots in Wukung some red-leafed cottons and were told they had pilose and other types with various morphological characters that they were using in an attempt to develop resistance to aphids.

Some land is prepared with tractors, but most of it is prepared with mules, oxen, and water buffalo. All cotton is hand-harvested and sold as seed cotton to gins owned by the state. Processing, compressing, and marketing are handled by the state. Markets are strictly controlled.

The planting seed are obtained from a regional Agricultural Research Institute where variety trials are conducted each year. The best-performing variety is selected and seed is provided to the communes, where it is increased in seed-production blocks. Acreage planted to cotton is strictly controlled by the government, allotments being given to each cotton-producing commune or state farm. The farm then may decide how much cotton each production brigade is to plant; the decision depends on the preference of the people, labor, soil types, availability of water, mechanization, and so on.

On our visit to the Tangwan Commune, a large cotton commune near Shanghai, we saw farms that we considered typical of commercial

cotton-growing farms in China. This commune, established in 1958, has
11 production brigades subdivided into 120 production teams. The labor
force numbers 12,000 people, who come from 5,400 households; the total
communal population is 23,000. The commune has 1,787 hectares of farmland. The major crops are cotton, rice, and wheat. Sixty percent of
the land is planted to grain, 30% to cotton, and 10% to vegetables and
medicinal herbs. The commune also produces pigs, chickens, and fish,
and it has a small dairy. In addition, there are nine small factories
that produce handtools for farming, furniture for the commune, small
machine tools, straw baskets, and mats.

The total crop yield per hectare on this commune was reported to be
10.51 tonnes, or about twice what the land was producing before the commune was formed. The commune has 501.6 hectares (7,600 mu) of cotton
planted in fields that usually range from 1 to 5 hectares in size. The
cotton yield in 1974 was 905 kg of lint per hectare. This was sold as
seed cotton. The first-quality cotton brought 52 yuan per 50 kilos.
Cotton is sold at the gin and is graded and classified by specially
trained persons employed by the state.

This commune is trying to become more mechanized. At present it has
only 23 tractors and 102 hand tractors. About 95% of the land is plowed
by tractors and the remainder is plowed with water buffalo. The cotton
and rice are transplanted from seedling beds to the fields by hand.
Cotton is planted from mid-March to mid-April and is hand-harvested
from late September to early November. The land then is planted to
rape, vegetables, or wheat.

The commune has five plant-protection specialists and four forecasting points. These specialists are primarily concerned with the control
of insect pests of cotton and rice. This group had received special
training in plant protection from the Shanghai Academy of Agricultural
Sciences and the Shanghai Institute of Entomology. They in turn trained
people in the production brigades and teams to assist in pest detection
and prognosis. In all, 380 people were involved in plant protection.
They use blacklight traps to detect insect flights and make field surveys
to determine the fields that need insecticide treatments. The pink
cotton bollworm and the cotton aphid are the main pests. Workers spray
three to five times with carbaryl for the pink bollworm. When aphids
appear, they may spray two or three times with dimethoate, which generally is added to the pink bollworm sprays. Sprays are applied with 3-
gallon compressed-air backpack sprayers, or sometimes with barrel
sprayers on wheels when the plants are small. Control measures also are
applied to reduce overwintering pink bollworms in the seed-cotton storage area.

Diseases of Cotton

Reports were received from several areas that the major cotton diseases
are "wilt" (both *Fusarium* and *Verticillium*), anthracnose, and *Phythium*.
Boll rots are also a problem in some areas where humidity is high in
the period when late bolls are formed.

Arthropod Pests of Cotton

The principal and minor arthropod pests of cotton in China are listed in Table 1, with the insecticides used for their control. The pink cotton bollworm and the cotton aphid appeared to be pests in every area we visited, with the former being the more serious of the two. Two *Heliothis* species were serious pests in Sian and Hunan but not in Shanghai. *Heliothis armigera* (Hübner) is probably the most important insect pest of cotton in China, and the pink bollworm is probably second. However, major efforts are being made to control the pink bollworm in seed-cotton storage areas, with the result that it is not as important as it once was. Economic thresholds used for initiating insecticidal treatment against certain of these pests are listed in Table 2.

TABLE 1 Principal and Minor Arthropod Pests of Cotton in China and Insecticides Used for Their Control

Pest	Average No. Treatments/ Season	Insecticide Used
Principal Pests		
Old World Bollworm *Heliothis armigera* (Hübner)	0-8	DDT, carbaryl, BHC, methyl parathion, dichlorvos, monocrotophos, chlordimeform, C-9140, phoxim
Pink bollworm *Pectinophora gossypiella* (Saunders)	0-4	carbaryl, C-9140, methyl parathion, chlordimeform, BHC (methyl bromide is used to fumigate stored cotton-seed for overwintering larvae)
Cotton aphid *Aphis gossypii* Glover	0-6	dimethoate, phorate, demeton, carbaryl, dichlorvos, monocrotophos, phosphamidon, naled, methyl parathion
Minor Pests		
Black cutworm *Agrotis ypsilon* Rottemberg	0-1	phorate
Spiny bollworm *Earias insulana* Boisduval		None given
Diamond bollworm *Heliothis asfulta* Guenee	0-8	Same as for pink bollworm
Spider mites *Tetranychus* spp.	?	tetradifon
Plant bugs *Lygus* spp.	?	BHC, DDT, methyl parathion, dimethoate

TABLE 2 Economic Thresholds for Initiating Insecticidal Treatment of Certain Insect Pests of Cotton in China

Pests	Treatment level
Bollworms, *Heliothis* sp.	15 eggs/100 plants
cotton aphid	(a) 150 aphids/100 plants
	(b) 3% of plants with curled leaves
	(c) 20% of plants aphid infested
Pink bollworm	10% of bolls infested

Measures Used to Control Arthropod Pests of Cotton

Entomologists and plant protection specialists in the communes all were placing emphasis on the integrated control of arthropod pests of cotton. They appeared to have good knowledge of U.S. pest-control practices and were well versed in the current jargon of the integrated control specialists.

Plant protection in China has high priority but must be applied in the context of the total agricultural production system. This includes management of crop protection times and sequences, irrigation, fertility, availability of machinery, elimination of pest breeding and hibernation sites, and crop rotation. All activities are directed toward a "good harvest," i.e., high production per unit of land. Integrated control fits well into this philosophy. Emphasis is given to environmental manipulation to suppress pest attack and insecticide treatments to prevent outbreaks. Some examples of integrated control that we observed, or were told about, are as follows:

Spiny bollworm Entomologists at the Peking Institute of Zoology reported that the spiny bollworm was a problem on cotton in southern China before the Revolution when cotton was planted in spring, summer, and fall. With the advent of the commune system, planting dates were regulated so that only spring planting was allowed. Immediately after harvest, stalks were destroyed and the fields were plowed and replanted with small grains. This combination of practices has reduced the spiny bollworm to minor pest status.

Cotton aphid Entomologists at the Peking Institute of Zoology reported that lady beetles were used for control of the cotton aphid during the past 3 years on more than 19,800 hectares of cotton. The beetles are hand-collected from various hosts and taken to the cotton fields for release. In addition, strip cropping of cotton with wheat is practiced, and as the wheat is harvested the lady beetles transfer to the cotton at a time when aphids may be a problem.

In Wukung, Shensi, entomologists at the Northwest Academy of Agricultural Sciences reported that before the Revolution cotton aphids were a serious pest in the area and were controlled mainly by insecticides. Control was excellent, but the chemicals killed the natural enemies, the aphids resurged, and secondary outbreaks of *Heliothis* became common. The aphids also developed resistance to many of the insecticides, so more toxic materials had to be used. There were also instances of poisoning of the workers. The Chinese now use insecticides selectively in combination with other measures, including winter destruction of the alternate hosts of the aphids. Cotton seed are treated with phorate for aphid control on seedling plants. After transplantation, they may send workers through the fields to remove by hand all heavily infested plants. In other cases, they first may spray with insecticides, such as dimethoate, only the individual plants that show signs of aphid infestation. If the aphids spread across the field, they may treat all the plants. In addition, they allow certain weeds to grow in the cotton in early season in an attempt to increase lady beetle populations. When the aphids appear, the weeds are pulled, forcing the lady beetles to the cotton. They also reported that they had developed a variety of cotton that had some resistance to the cotton aphid. This variety is planted over a large part of the province. (We requested but were unable to obtain seed of this variety.) Resistance to the aphid is due to increased pubescence, which makes it less attractive to the pests. It also possesses some antibiosis; the aphids multiply at much slower rates on the resistant variety than on the nonresistant varieties.

Pink bollworm Principal control measures are directed against the potential overwintering (diapausing) larvae of the pink bollworm. When the cotton is harvested, it may be spread to dry in the sun. The heat drives out many pink bollworm larvae, which may be consumed by chickens or killed by hand. Rotten and heavily infested bolls left in the field are picked by hand. In addition, chickens may be placed in the cotton fields to eat any seed left after harvest. Stalks are cut and used for winter fuel in the workers' homes. The fields are plowed, planted to a winter crop, and irrigated. This kills many larvae and forces the premature emergence of many moths in the spring before the cotton is susceptible to attack.

The storage places for seed cotton are sprayed with insecticides such as BHC or DDT before the cotton is brought in. Larvae leaving the cotton to hibernate in the crevices of the walls are killed by the insecticide. The seed may often be stored on straw mats where some larvae may spin hibernaculums or go to pupate. These are killed by hand. In the cold climates the storerooms are allowed to become quite cold (-6° C), and this practice freezes the larvae. It is also a general practice to release the pink bollworm parasite, *Dibrachys cavus* Walker, in the storerooms. Larvae mortalities of 75%-90% were reported. Seed also may be fumigated with methyl bromide.

In the growing season, blacklight traps are used to capture moths and to assist in making decisions on insecticide treatments. Treatments are applied as needed to prevent outbreaks.

Other Bollworms Cotton in China is attacked by two species of bollworms, the Old World bollworm and the diamond bollworm. Of the two
species, the Old World bollworm is by far the more serious pest of cotton in China. In Shensi, the crop may be treated seven or eight times
with insecticides for control of the Old World bollworm. Control measures are initiated through the system of insect forecasting, moth
catches in blacklight traps, and field infestation records. Field records are made at three or four points in each commune. Once 15 or more
eggs per 100 plants are found, insecticides are applied. Usually, the
aim is to obtain control before damage to the crop can occur. However,
this system does not provide insect natural enemies with much opportunity to control the bollworms and leads to the use of more insecticides
than would be needed if control were based on a combination of egg and
larval numbers plus small amounts of damage to squares and bolls.

In the Shanghai area, we were told that *Heliothis* bollworms are not
a problem, probably because of the cool winter weather. However, the
cotton experts on our team disagree with this hypothesis. Shanghai is
on nearly the same parallel as New Orleans and has a climate similar
to U.S. areas where bollworm infestations are extremely serious. The
more likely cause is the fact that the land is multiple-cropped, being
plowed at least twice and irrigated one or more times between the harvest of one cotton crop and the planting of the next. Also, no wild
alternative hosts are allowed to grow, and few cultivated host plants
(other than cotton) are produced. Thus, the fall, winter and spring
tillage and irrigation between cotton crops probably have more to do
with keeping *Heliothis* numbers at low levels than the winter weather.

Cotton leafworm The cotton leafworm (genus unknown, but probably
Prodenia spp.) was reported by the Peking Institute of Zoology as being
a major pest of cotton in much of southern China. Officials there reported that in limited tests an NPV virus of this pest has provided good
field control--75%-100% mortality. No other information was provided
on this pest.

Plant bugs *Lygus* were mentioned several times as pests of some consequence but no information was provided on economic thresholds or summer
spray treatments. Dimethoate apparently is the insecticide most often
used on these pests.

Red Spider mites These were mentioned as being occasional pests on cotton but almost nothing was said about their control.

REFERENCES

1. Efimov, A. L. and G. M. Miftakhov. 1954. The pink bollworm and
    other cotton pests in China. Zool. Abur. 33(5):1065-1079.
2. Kung, Peter. 1975. Farm crops of China. World Crops. (May/June).
    pp. 123-132.

C. GRAIN

General

The U.S. Department of Agriculture[1] has estimated that in 1974 China produced 120 million metric tons of rice, 32.5 million of wheat, and 65.2 million of other grains, including corn, sorghum, barley, millet, rye, and oats (USDA, 1976). Although rice is the most important grain crop, the others combined provide more tonnage and calories. So insect control on these crops is an important consideration.

Some of these grains as well as soybeans are used as parts of the intercropping system so well utilized in China. In the following diagram, crops connected by a line are intercropped:

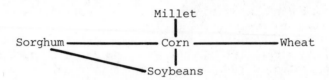

No formal statistics on intercropping as an insect-control measure came to our attention in China. Research results obtained by the International Rice Research Institute[2] in the Philippines, however, show that corn borer infestation in corn may be reduced by intercropping with peanuts, and the overall land-equivalent ratio of intercropping versus solid corn planting in the Philippines is 1.3 to 1.6 (IRRI, 1972). There is a corresponding increase in labor input. Thus, intercropping can be a rewarding system in situation where land rather than labor is the chief limiting factor.

Winter wheat is a versatile crop for use in multiple-cropping sequences, with corn, soybeans, rice, and cotton. Corn is relay-planted among the rows of wheat, and the corn is later followed by a green manure crop. Corn may also be relay-planted in an earlier planting of corn that is then followed by wheat. The U.S. Department of Agriculture estimates that on the national average, China produces 1.4 crops annually by multiple cropping; i.e., an acre of land produced, on the national average, 1.4 crops annually due to the use of multiple cropping (USDA, 1976).

These concentrated systems, centered entirely on grain crops (wheat, corn, sorghum, millet, and other grains), often suffer from and perpetuate many of the same insects, including many soil forms.

Presentations by scientists from several institutions visited dealt with some of these insects. Recent published reports have also been consulted. Information from these sources is incorporated here.

Wheat Insects

Numerous insect (and mite) pests of wheat are listed in two recent publications and referred to by Chinese scientists. They are as follows:

Chewing forms

Mythimna separata (Walker)	Armyworm	(3)
Meromyza saltatrix Linnaeus	Wheat stem maggot, yellow	(4)
Oscinella pusilla	Wheat stem maggot, black	(3)
Nanna truncata Fan	Wheat shoot maggot	(3)
Dolerus tritici Chu	Wheat sawfly, small	(3, 4)
Pachynematus sp.	Wheat sawfly, yellow	(3)
Dolerus hordei Rhower	Wheat sawfly, large	(3)
Ochsenheimeria sp.	Wheat stem borer	(3)

Sucking forms

Sitodiplosis mosellana (Gehin)	Wheat midge, red	(3, 5)
Contarinia tritici Kirby	Wheat midge, yellow	(3, 5)
Haplothrips tritici	Thrips	(3)
Aelia acuminata Linnaeus	Plant bug	(3)
Macrosiphum granarium Kirby (11a)	Aphid	(3, 4)
Rhopalosiphum maidis (11a)	Aphid	(4)
Rhopalosiphum padi Linnaeus (11a)	Aphid	(4)
Toxoptera graminum (Rhondani) (11a)	Aphid	(3, 4)
Myzus persicae Sulzer	Aphid	(4)
Rhopalosiphum erysimi Kaltenbach	Aphid	(4)
Thamnotettix oryzae Mats (11b)	Leafhopper, rice	(3)
Deltocephalus striatus Linnaeus (11b)	Leafhopper, mottled	(3)
Thamnotettix sp. (11b)	Leafhopper, black	(3)
Cimotettix triatus (11c) (11d)		(5)
Delphacades striatella Fallen (11e)	Planthopper, grey	(5)
(11f)		(3)
Penthaleus sp.	Mite, round	(3, 4)
Petrobia latens (Muller)	Mite, long legged	(3)
Aceria tullicae Linnaeus (11g)	Mite	(3)

Soil insects

Gryllotalpa africana Palisot de Beauvois	Mole cricket	(3, 4)
Gryllotalpa unispina Saussure	Mole cricket	(3)
Holotrichia diomphalia Bates	Scarabaeid	(3)
Holotrichia morosa Waterhouse	Scarabaeid	(3)
Holotrichia titanis Reitt	Scarabaeid	(3)
Anomala corpulenta Mots	Scarabaeid	(3)
Anomala sp.	Scarabaeid	(3)
Pentodon patrualis Frivalsky	Scarabaeid	(3)
Trematodes tenebroides Pallas	Scarabaeid	(3)
Agriotes fusciollis Miwa	Wireworm	(3)
Pleonomus canaliculatus Faldermann	Wireworm	(3)
Melanotus caudex Lewis	Wireworm	(3)
Apophilia thalassina Faldermann	Leaf beetle	(3)

Wheat insects were discussed at the College of Agriculture, Wukung, Shensi, and at the Kirin Academy of Agricultural Sciences, Sian.

Midges and Stem Maggots These insects had been serious pests previously but are no longer problems. Breeding programs in 1950-1956 produced resistant varieties which reduced their populations to below economic levels. The basic germ plasm from Chung-mung No. 28 and Pi-ma both showed resistance in the field. The cross of the two lines produced 6028 which is resistant. Different crosses of 6028 were made. Backcross with Pi-ma produced Shantung No. 9, a cross with Shan-mai No. 1 produced Fung-sho No. 3, and a cross with Ah-pei produced Kuan-chung No. 1. The progeny of these crosses are all resistant, and are generally planted. Besides these, Chi-li (Kirin Success) and Ai-kan-hung (Dwarf Red) are commonly used (3).

Toxoptera graminum *Toxoptera graminum* is the most important vector of barley yellow dwarf in Shensi, although three other aphids can spread the disease. It can migrate long distances, and develop at low temperatures. It infests millet in the summer and migrates to wheat in the fall, where it propagates and overwinters. It prefers fields with full exposure to the sun, thus, such fields are the target for monitoring. This involves determining the time of the three- or four-leaf stage of wheat and measuring the population density of aphids at that time, the percentage of virus infection (through bioassay in the greenhouse), and the rate of virus propagation. This information gives the potential for infestation and infection the next year.

In the spring, the following factors are monitored:

- Population density of the aphids.
- Percentage of virus infection.
- Rate of propagation and dispersal of the aphids.
- Weather condition, especially any decrease in temperature during and after heading. Temperatures below 30° C favor virus infection.
- Abundance of natural enemies, e.g., lady beetles and braconids.
- Wheat variety. Resistant varieties are Fung-sho No. 3 and No. 132.

The need for control is determined on the basis of the above information. In ordinary years, large-scale control is not needed.

Mythimna separata (*Walker*) This armyworm is very destructive. It feeds on leaves and stems of wheat and on other gramineous crops used in the multiple-cropping systems. Scientists at the Plant Protection Institute, Kirin Academy of Agricultural Sciences, determined that the insect does not overwinter in that area, so a cooperative project was organized to study its migration. They marked and released 600,000 moths in Shantung, Anhwei, and Kiangsu provinces. Several hundred checkpoints were set up in Liaoning and Kirin provinces. Some 2,600,000 moths were collected; these included six marked individuals in Liaoning and three in Kirin. The study showed that moths may migrate from as far south as 2,000 km. Subsequent studies established the following:

- The insect has 6 to 8 generations a year in southern China, 4 or 5 in central and eastern China, 3 or 4 in northern China, and 2 or 3 in northeastern China.
- South of the 27th parallel (Kwangtung and Fukien provinces), the first-generation population infests wheat, and adults appear in middle and late March. The adults migrate northward to the area between the 30th and 36th parallels, where they continue to multiply.
- They infest wheat in the last-named area in April and May. When a generation is completed there in May or June, the adults move northward to Inner Mongolia, Hopei, northern Shensi, and northern Shansi, where they reproduce.
- There they infest wheat, corn, millet, and other grains and reach adult stage in middle and late July. The adults then move southward to Hopei, Shantung, Honan, and eastern Shansi.
- In these last-named areas, they infest millet and corn and reach the adult stage between mid-August and late September. The moths commonly move southward to south of the Yangtze River.
- In the south, they infest late rice and winter wheat. Adults emerge the next spring and start the cycle again.

Thus, it is established that information on the population density in the south during February and March will foretell the potential of infestations in the north.

Insect-pest monitoring and forecasting are conducted at numerous stations organized at five levels: province, region, county, commune, and brigade. In Kirin, there are 186 networks, each consisting of many stations. Uniform methods of monitoring are developed through discussion among farmers and technicians. On the basis of long-term records of population trends, damage relationship, and the current monitoring results, forecasting criteria are developed.

Information on insect status is mailed or cabled (with special coding to the upper level of the organization), and recommendations are issued. Summary reports are published every 10 days during the summer months. Records, mailed and cabled material, and summary reports of 1974 were on display at the Forecasting Laboratory, Plant Protection Institute, Wukung, Kirin.

The forecasting networks in Kirin are also involved in monitoring the occurrence, abundance, and development of aphids, the European corn borer, and other target species.

Soil insects Soil insects, such as mole crickets and wireworms, are serious pests in many areas, including Kirin and Shensi provinces. These insects are controlled with insecticides in corn fields (see "Corn Insects").

Corn Insects

Corn, planted in many parts of China, is attacked chiefly by the European corn borer. Many general feeders, such as locusts, armyworms, and wireworms, also attack corn.

Wanting to ascertain whether the Chinese form of the European corn borer is the same as the form in the United States, we brought a dozen Minnesota borer moths, preserved in 70% ethanol, to Peking. These were presented to the Institute of Zoology, Academia Sinica, and were found to be the same as the Chinese form.

This information is important because the species infests not only corn and other grains but also cotton, legumes, and vegetables in China. In southern China, the insect is as injurious on cotton as on corn. It sometimes attacks cotton in the United States, but infestations are seldom serious.

Control practices used for the European corn borer in five locations were observed and discussed--or extracted from literature--and are described here.

First location In Kungchuling, Kirin Province, research on control of the insect involves four aspects. Integrated control is the common approach, and use of *Beauveria*, *Trichogramma,* and pheromones is being studied.

Beauveria is produced locally by the farmers with simple materials (wheat bran, wood powder, grass cuttings, etc.) and cultured in simple containers (trays, plastic bags, pans, and outdoor pits). A granular formulation is prepared with cinders or sand as a base. When *Beauveria* is applied on piled infested cornstalks, it kills the overwintering corn borer larvae, especially when they start to move in the spring. This practice commonly reduces the field moth population by 75%. *Beauveria* is also applied to first-brood larvae, and here 82% kill is achieved. Infected larvae produce spores that infect the second brood.

In one field, we saw many corn borer larvae and corn earworm larvae that had been killed by *Beauveria*. We saw only one live corn borer larva.

In the Tung Hua region, three releases of 20,000 *Trichogramma* wasps per mu (120,000 per acre) produced 48%-95% corn borer mortality. In 4 years of *Trichogramma* releases, the number of borers per 100 plants dropped from 214 to fewer than 20.

Habits of the European corn borer were observed to determine the usefulness of pheromone in monitoring and controlling it. The moths mate after midnight. The first-brood moths find shelter in grain fields, and the second brood in soybean fields. These are the best places to catch moths of the two broods.

In one-brood areas, *Trichogramma* are usually sufficiently effective without use of chemicals, but *Beauveria* is commonly used as a supplement. In two-brood areas, the main control method is *Beauveria* to kill overwintering larvae, followed by use of the chemical carbaryl. If the attempt to synthesize the pheromone and use it as a direct control measure or for monitoring is successful, it is expected that it will be used in combination with *Beauveria* and insecticides.

Second location At the Dayushu (Big Elm Tree) Brigade of Nanweitse People's Commune (20 km from Kungchuling), *Beauveria* has been used to control the European corn borer since 1973. It is applied in granular form at the late tassel stage in early June at a dosage of one larval

equivalent on 250 plants. Fall treatment of cornstalk piles is also
common. *Beauveria* production is said to be simple, and its effects
are said to be long lasting. Since 1973, its use has reduced the corn
borer damage rate from 50% - 60% to 15% - 18% and the number of borers
per 100 plants from 30 to 6 or 7. No bad effect on workers has been
noted. The insecticide carbaryl is also used.

Third location In Sian, Shensi Province, a demonstration field of corn
planted by second- and third-year students of the College of Agriculture
was visited. The variety was a yellow-white single cross. The yield
is about 7,500 kg per hectare, whereas the yield of local varieties is
between 4,500 and 5,200 kg. It is resistant to leaf blight and has
good adaptability. One disadvantage is the very tall stalk. Growth
in the field looked good and was free of signs of insect infestation.
It was indicated that four management strategies are emphasized by
the College of Agriculture:

- Use of a good variety.
- Deep plowing (40 cm).
- High-density planting--45,000 plants per hectare. This is equivalent to 18,000 plants per acre, a rather low density in the United States.
- Proper fertilization and irrigation. During the seedling stage,
fertilizer is applied at a low rate to reduce disease infection. Later,
the crop should be fertilized twice--each application should include
6,000 kg of manure per mu (90 metric tons per hectare)--and irrigated
twice.

In the demonstration field, the following chemical control program
was used:

- Soil treatment with BHC or dimethoate on an annual routine basis
for wireworms.
    - A bait consisting of alfalfa and trichlorfon for mole crickets.
    - BHC applied at tasseling to kill thrips.
    - BHC and granular trichlorfon to control the European corn borer.

Fourth location In the northern part of Shensi Province, a single cropping of corn is planted in the spring. The European corn borer is a
problem here. In the central part of the province, however, corn is
planted in early summer following wheat. Here the European corn borer
is not a problem, because the corn is then grown too late in the season
for oviposition by the moths issuing from the overwintered larvae.
A mass crop-utilization method, shredding stalks for animal feed,
destroys corn borer larvae and is advised as a control.

Fifth location The fifth location was Honan Province. The delegation
did not visit this area, where the European corn borer has three generations a year. A 1975 report (Agricultural Bureau, 1975) indicates that
in Honan cultural, physical, biological, and chemical control measures
were equally emphasized in an integrated control approach. Two aspects,
the use of *Trichogramma* and the use of blacklight traps, are summarized
below.

   Trichogramma were released in 1974 in about 5,000 hectares (75,000
mu) in Lai-Kou County, 44,000 of which were for corn borer control. Two
releases of about 150,000 per hectare (10,000 per mu) during the first
brood and two more during the second brood resulted in parasitism of
68.2% and 78%. The percentage of damaged plants was reduced from 63%
to 10% for the first brood and from 31% to 11% for the second brood.
The infestation by the third brood was greatly reduced in fields where
Trichogramma had been released for the second-brood borers.
   In the same area, 8,000 blacklight traps were set up in about 20,000
hectares (300,000 mu), and they trapped more than 33,000 kg (66,000 jin)
of insects in 1974. The traps were especially effective for corn borer
moths, catching up to 2,000 per night. Random checking of 2,032 moths
showed 57.5% to be females, and 76.7% of these contained eggs. The overall oviposition rate was reduced 55% - 60%, and the number of infested
plants was reduced by 50%. The use of insecticides was also reduced,
resulting in a saving of 67% - 90% in cost of materials and 77% - 80%
in cost of labor.

Other Grain Insects

Sorghum and millet were seen in solid plantings and in interplantings
with corn. Inspection of these fields yielded no information on insect
infestation. No presentation during the trip dealt with these crops.
However, a Shantung plant-protection handbook[8] lists the sorghum aphid,
the striped sorghum borer, and various other borers as pests of the
panicles. Included are the grain borer, peach borer, cotton bollworm,
and corn borer. No scientific names were given for these insects.
The sorghum aphid was reported to cause the most serious damage when
abundant.
   A recent report (1975) indicates that sorghum fly infestation in some
fields reached 20%[9] (Utilization of Crop Hybrid Dominance Cooperation,
1975). In 1972, when some varieties had 40% dead shoots, variety 3197A
was not affected.
   Scientists at the Plant Protection Institute in Kirin commented
briefly on the sorghum aphid. Knowledge of its overwintering hosts are
important because the population on such hosts will provide forecasting
information for the sorghum infestation.
   The Shantung plant-protection handbook lists the millet stem borer
and the millet panicle worm as pests of millet. The former enters the
stem at ground level and may cause death of seedlings and "white head"
or "broken stem" on growing plants. A recent report (1975)[10] indicates
that in the 1950's scarabaeids were very serious pests cf millet. But
BHC and deep plowing were so effective that these insects have been
kept at low densities.
   In the 1960's, the millet stem borer became important. The following
procedures were developed for its control: Use of a specially designed
scraper that scrapes up the stubbles harboring the borers, delaying
planting so the plants will escape heavy oviposition, and planting an
early crop to concentrate oviposition. The trap crop is then treated
with BHC following peak oviposition.

A report also mentions that since 1970 aphids and leafhoppers have increased their importance as vectors of red leaf virus and dwarf virus, diseases of millet.

REFERENCES

1. USDA. 1974. Agriculture in the United States and the People's Republic of China, 1967-71. Foreign Agricultural Economic Report No. 94 (Cited in US NAS Plant Study Trip Report).
2. IRRI. 1972. International Rice Research Institute Annual Report. pp. 21-34. Los Banos, Philippines.
3. Publication No. 13. "Pictures of Chinese Crop Diseases and Insect Pests" Editorial Committee. 1972. Diseases and Insects of Wheat, Volume 2 of "Pictures of Chinese Crop Diseases and Insect Pests." Agricultural Press, Peking. 86 pp. including 42 color plates, with scientific names (in Latin) and Chinese names of insects and pathogens. Text in Chinese.
4. Publication No. 19. "Handbook for Plant Protection Workers" Editorial Committee. 1974. Handbook for Plant Protection Workers. Composite edition. Shanghai People's Press, Shanghai.
5. Presented by scientists at Shensi Provincial College of Agriculture and Forestry. Wukung, Shensi.
6. Publication No. 32. Northwest Institute of Agriculture. 1956. Research and control of wheat midges. Shensi People's Press. 75 pp. plus 2 color plates. Scientific names (In Latin) of insects. Text in Chinese.
7. Agricultural Bureau, Revolutionary Committee of Lai-Kou Hsien, Honan Province. 1975. The integrated control of European corn borer. Acta Entomol. Sinica *18*:156-160.
8. Publication No. 21. Shantung Provincial Revolutionary Committee Agriculture Bureau. 1972. Control Methods of Cereal Diseases and Insects. Shantung People's Press, Shangtung. 141 pp. plus 26 colored plates. Text in Chinese.
9. Utilization of Crop Hybrid Dominance Cooperation, Kwangtung Province; Science and Technique Bureau of Taisan County, Kwangtung Province. 1975. A preliminary study on the sorghum fly, *Atherigona soccata* Rondani. Acta Entomol. Sinica *18*:47-51.
10. Research Group of Dai-Bau Production Brigade Fan-Sy Hsien, Shansi Province. 1975. Integrated control of insects injurious to millet. Acta Entomol. Sinica *18*:301-306.
11. Virus diseases transmitted by various insect vectors:
    a) Barley yellow dwarf virus
    b) Wheat red dwarf virus
    c) Wheat blue dwarf virus
    d) Winter wheat mosaic virus
    e) Northern cereal mosaic virus
    f) Wheat black dwarf virus
    g) Wheat streak mosaic virus

D. SOYBEANS AND PEANUTS

In China soybeans are used as a fresh vegetable (green beans), as dry beans, and in a variety of processed foods (soy sauce, soy milk, bean curd products, and bean oil). The average annual production of beans for 1971-1973 was 10,283,000 tonnes, which was 21.2% of the world production[1]. Peanuts, another important legume crop, are used as food and for oil.

Soybeans

Northeastern China is the most important bean-production area in the country. We saw extensive culture of this crop in Kirin Province. Discussion at the Kirin Academy of Agricultural Sciences on soybean insects involved mainly the soybean pod borer, *Grapholitha glycinivorella* Matsumura, which is the most injurious soybean pest there. The soybean pod borer was listed as *Encorma glycinivorella* in the report on the U.S. plant studies trip in 1975.

Control methods for the pod borer are described below.

Cultural control Resistant varieties (e.g., Kirin No. 3) are used. Very young soybean pod borers cannot penetrate the pods of this variety. Wheat is usually planted following soybeans. When the wheat is harvested, the bean pod borers pupate in the soil. Plowing at that time kills about 80% of pod borer pupae, and the subsequent populations are correspondingly reduced.

Chemical control Use of insecticide only with heavy infestations and only selectively, on the basis of forecasting, is recommended. Sorghum stems should be soaked in dichlorvos and then implanted or pressed into the soil to give a fumigating action. Thirty stems per mu (about 450 per hectare) were effective, and the method does not affect natural enemies of the borers.

Natural enemies Ichneumonids and braconids cause about 60% parasitism. In Kirin, about 80,000 hectares (1,200,000 mu) were planted to soybeans in 1975. About 60,000 hectares (900,000 mu) were planted to varieties resistant to the soybean pod borer and about 3,000 hectares (200,000 mu) were planted to wheat and plowed under. With this program, rate of damage was kept at about 10%, and quality was good.

Possible use of *Trichogramma* and *Beauveria* is under study.

A brief mention was made of aphids on soybeans. Knowledge of the overwintering hosts is important, because populations on such hosts will provide forecasting possibilities for infestations on soybeans.

Many other species of insects are pests of soybeans and other beans in China. They are listed in the *Handbook for Plant Protection Workers*[2] as follows:

Etiella zinckenella (Treitschke)
Maruca testulalis Geyer

[1]pea-pod borer
[2]soybean borer

Heaylepta indica Fabricius	[2]soybean leafroller
(*Lamprosema indica* Fabricius)	
Clanis bilineata Walker	[3]bean hawkmoth
Dasychira locuple Walker	[4]soybean tussock moth
(*Cifuna locuple* Walker)	
Epicauta gorhami Marseul	[4]bean blister beetle
Holotrichia gebleri Faldermann	[5]southern large black scarabaeid
Holotrichia diomphalia Bates	[5]northern large black scarabaeid
(*Lachnosterna diomphalia* Bates)	
Anomala corpulenta Motschulsky	[5]bronze scarabaeid
Maladera orientalis Motschulsky	[5]black velvet scarabaeid
(*Serica orientalis* Motschulsky)	
Aphis craccivora Koch	[6]bean shoot aphid
(*A. medicaginis* Koch)	

Additional insects are discussed in *Control Methods of Cereal Diseases and Insects*[3] but only in Chinese. They are the bean inchworm, bean stem maggot, bean aphid, and bean weevil.

Peanuts

The delegation was not able to visit any of the major peanut-producing areas of China, but we saw small fields of peanuts from the car in every province that we visited. Roasted peanuts of excellent quality, both Virginia runner and Spanish types, were served at several meals. Canned peanuts also were available in every store, market, and hotel that we visited, indicating that peanuts are a favored food in China. The peanut flower also is used as a nitrogenous ingredient in the *B.t.* cultures made at some communes.

The only peanut field that we were able to visit was on the experimental farm of the Institute for Plant Protection for Hunan Province near Changsha. The peanuts were of a Spanish type (nonrunner) and heavily loaded with nuts. The crop was being produced under irrigation and with a high level of fertilization. We were told that the major pests were the cutworm, *Agrotis ypsilon* Rottemberg, and the army worm, *Prodenia litura* Fabricius. When treatment is required, they use dimethoate or trichlorfon. They said they did not have any problems with soil insects such as rootworms or the lesser cornstalk borer. The peanuts were not being grown for experimental purposes but were for use by the people on the farm. We left them peanut seed from Texas and asked for seed in return but were told that none was available. We did collect a few full-sized green nuts with the hope that they might be mature enough to germinate.

REFERENCES

1. FAO Agricultural Production, 1973, Rand.
2. "Handbook for Plant Protection Workers" Editorial Committee. 1974.

58

Handbook for Plant Protection Workers. Composite edition.
Shanghai People's Press, Shanghai.
3.   Shangtung Provincial Revolutionary Committee Agriculture Bureau.
1972. Control Methods of Cereal Diseases and Insects. Shangtung
People's Press, Shantung. 141 pp. plus 26 color plates. Text in
Chinese.

E.   CITRUS FRUITS

Cultivation of citrus in China dates to at least 2286 B.C. The crop
has been produced commercially since about the tenth century A.D.[1]
(Burke, 1967). Most species of citrus are native to China (and South-
east Asia). This fact made the crop of particular interest to the
delegation because citrus is exotic to the United States but is cultivated extensively in Florida, California, Texas, and Arizona.

As is often the case, some citrus pests were imported with their hosts
into the United States without the natural enemies that had adapted to
the citrus insect pests in their native lands. Some beneficial species,
largely parasites of scale species, were imported from China many years
ago. But most entomologists who are responsible for citrus-plant protection programs agree that additional importation of beneficial species is desirable. For example, parasitism of the California red scale,
Aonidiella aurantii Maskell, in California is generally good on coastal
areas, but effectiveness declines to very poor in the hot, arid interior
valleys. The need for parasites that would regulate the population of
the red scale in these latter areas is obvious.

In view of this interest, the delegation requested an opportunity to
inspect and discuss citrus with their host counterparts, which was provided while visiting the Changsha area of Hunan Province at the Institute of Plant Protection. Discussions on citrus-plant protection were
quite limited, but a brief summary, excluding most of the biological
control observations, is included.

Information on total production of citrus in China was not available.
In 1959, the National Chinese News Agency reported total plantings as
97,500 hectares and total production as 12 million boxes, each weigh-
ing 31.8 kg. Total commercial citrus plantings in 1964 on a solid-
planting basis may be estimated as not exceeding 39,000 hectares, but
plantings were increasing[1]. Most of the southern provinces of China
produce this crop in varying amounts, but Kwangtung, Kwangsi, Fukien,
Hunan, and Szechwan have the largest number of trees. A number of
citrus species and varieties are grown in these provinces. Several
groves in Hunan Province were examined, but time permitted intensive
inspection of only one. Also, a detailed lecture on control of citrus
pests was presented at the Hunan Institute of Plant Protection.

The main citrus produced in Hunan Province belong to the Mandarin and
sweet orange groups. Most of the groves observed were small and closely
planted. Quite commonly, annual crops (e.g., vegetables) are planted between the tree rows. Because of their nitrogen-fixing value, legumes
are preferred for interplanting, and soybeans were observed most frequently. Many of the groves are planted on gentle hillsides, usually

on small terraced areas, which are unsuited for rice production. It is
our understanding that many of the soils on the hillsides have a high
clay content, resulting in poor drainage, which often results in a short
productive life of the trees.

Pests of citrus are limiting factors in successful production, and
their control is important. There are, for example, as many as 15
plant diseases and 70 species of arthropod pests. According to the
lecturer, control was difficult prior to the Revolution because there
was little research or operational support. At present, however, teams
of trained technicians are supported by a research base that provides
the scientific knowledge and technology needed for successful control.

Prior to 1960, five insect species were dominant pest problems on
citrus in Hunan Province:

Icerya purchasi Maskell	cottony cushion scale
Ceroplastes rubens Maskell	red wax scale
Clitea metallica	citrus leaf beetle
Anoplophora chinensis (Forster)	white-spotted longicorn beetle
Contarinia sp.	citrus bud maggot

The control measures developed have reduced these species to less
than dominating status. As a consequence, they are now considered
secondary pests. The primary action measures used in the program include:

● Grove sanitation, e.g., pruning.
● Introduction of the "red lady beetle" (presumably this is the
vedalia, *Rodolia cardinalis*, a highly effective predator of the cottony
cushion scale).
● Removal and renewal of unthrifty and dying trees.
● Manual control of pests.
● Insecticide application.

Since the reduction of the five species listed above to secondary
status, six other species are now considered to be significant secondary or major pests:

Panonychus citri (McGregor)	citrus red mite
Phyllocoptruta oleivora (Ashmead)	citrus rust mite
Parlatoria pergandii (Comstock)	chaff scale
Chrysomphalus ficus Ashmead	brown round scale
(*C. aonidum*)	(Florida red scale)
Phyllocnistis citrella Stainton	citrus leaf miner
Nadezhdiella cantori	brown longicorn beetle

It was not clearly stated to the delegation whether current pest-
control practices, particularly insecticide use, had elevated the pest
status of these latter species to consistently damaging population
levels. Experience in other citrus-producing areas of the world has
demonstrated that this can occur on citrus, especially in the case of
the citrus red mite, some scale species, and leaf miners.

Evidence supporting faunistic changes in citrus groves following insecticide applications was available in Kwangtung Province.[2] The authors divided the evidence into four stages of occurrence: the stage of indigenous pesticide application (1935-1952), the stage of organochlorine pesticide use (1953-1958), the stage of organophosphorous pesticide use (1959-1964), and the stage of combining acaricides with other pesticides (1965 to the present).

In the second stage, many chewing and sucking species were eliminated as pests. In the third stage, additional pest species were eliminated, but pesticide resistance became a problem with some species, particularly the citrus red mite, which was elevated to serious pest status. To counter this, it was necessary to add acaricides to the pesticide program (fourth stage). Often the acaricides are combined with insecticides. Other problems then arose, but some are thought to be the result of changing production practices. For example, fruit-bearing noctuid moths are more serious, and this increase is believed to be related to planting citrus on hillsides where the moths are found more abundantly.

The authors ascribe much of this problem to "indulgent" use of pesticides (presumably overtreatment). Such use greatly reduced populations of natural enemies and, concomitantly, their effectiveness in regulating pest populations. In general, they found that fewer species could still be considered pests, but those surviving treatment became major concerns as their populations became very high. It was proposed that integrated control programs be instituted and that insecticides be used only as proved necessary; dosage rate and frequency of application were considered particularly important. Agricultural production practices are to be the fundamental control measure, and these practices are concentrating initially on evaluating the effectiveness of natural enemies (see biological control section).

In Hunan Province, the research and technical workers, along with the peasants, have made biological and some ecological studies of the major arthropod pests. The seasonal history and related needs for control were reviewed for 5 of the 6 major pests in Hunan Province.

Citrus red mite

The citrus red mite thrives best in warm, dry weather. There are 12 to 15 generations a year, with oviposition starting in mid-March. Populations peak in April and May. In May, unusually high temperatures and natural enemies reduce populations below damaging levels. In September and October, populations again increase. The principal control method consists of a winter spray but additional sprays are applied during the growing season when population densities are high as determined by population sampling and forecasting.

Citrus rust mite

The citrus rust mite has more than 20 generations a year in Hunan
Province. The species overwinters as an adult, and about mid-April the
adults move to the citrus buds, which are injured by their feeding. Migration onto the young fruit occurs between middle and late May. Population increase is rapid during June through August when temperatures
range from 20° to 30° C and monthly rainfall is less than 100 mm. When
these conditions prevail, chemical control is usually needed. The key
to successful control with chemicals is a winter spray followed by repeated sprays in mid-June, July, and August. A fungus disease of the
citrus rust mite has been observed.

Chaff scale

The chaff scale has three generations a year. The first-generation
nymphs (crawlers) appear in early May. Most of them settle on and injure branches; some settle on and injure fruit. The second generation
appears from early to mid-July and the third in early September. The
second and third generations cause major injury to the fruit. The
highest populations are found in the shaded parts of the trees. Control is by winter pruning and spraying in the early stages of nymphal
development.

Citrus leaf miner

The citrus leaf miner has more than 10 generations a year. The adults
of the first generation emerge in late May and the second in late June;
subsequent generations overlap. The adults oviposit in young leaves,
and damage is most serious from July into September. Damage is most
severe on growth following budding and on seedling and young trees.
Chemicals are applied to control the adults during peak occurrence.

Brown longicorn beetle

This cerambycid beetle has one generation every 2 years. The adults
emerge from early April to mid-September. Two population peaks occur--
the first from mid-April to late May and the second late in the season.
The eggs hatch in 7-16 days, the time depending on temperature, and the
larvae bore into the tree bark and on into the wood. The heaviest attack is at the base of the tree. Damage is most serious to old and de-
clining trees and is especially serious where a tree has received poor
horticultural management. The key elements in control are prevention
through good management and renewal of badly affected trees. Larvae
are often dug out, and populations can be reduced by catching the adults
in light traps and destroying them.

Additional Comment

The life history of the brown scale was not reviewed for the group. Other insect pests occasionally demand attention. For example, two species of tortricid leaf rollers, *Homona coffearia* and *Adoxophyes privatana,* can cause damage. Certain other species of interest to us did not occur in the area of Hunan Province visited. The California red scale, *Aonidiella aurantii,* for example, was said to occur farther west in Hunan Province.

Four points were emphasized concerning control of citrus pests in the province. In this program, an attempt is made to follow integrated control practices insofar as they have been developed.

Careful sampling A forecasting service, similar to the one for rice, is used. When control is needed, the workers in the communes are urged to observe well-designed programs.

Improvement of citrus grove management Good management includes sanitation practices, properly timed pruning, and use of proper amounts of fertilizer and irrigation water. Generally, all inputs that improve tree vigor and productivity are recommended.

Use of multiple techniques to destroy pests These include light traps for leaf miners, longicorn beetles, and leaf rollers; handpicking and destruction of beetle adults and larvae; reducing populations of *Agrilus* borers by scraping off the superficial layer of bark; and digging out young longicorn beetle larvae.

Proper application of insecticides Techniques of application have been developed, and use of insecticides is recommended only during periods of pest abundance.

- Mites Tetradifon, dimethoate, carbophenothion, lime-sulfur, and malathion are applied.
- Scales Malathion, carbophenothion, and pine resin-soda mixtures are applied.
- Leaf miners and leaf rollers Dimethoate, phosmet, and chlordimeform are applied.
- Longicorn beetle larvae Dimethoate and water solutions of dichlorvos are applied. Often cotton is dipped into the dichlorvos and the treated cotton is stuffed into the holes containing the beetle larvae.

In general, six or seven chemical applications are made each year. The highly toxic insecticides (e.g., parathion and demeton) and the long residual organochlorine compounds (e.g., DDT and BHC) are being phased out. Emphasis has been placed on alternative application of different chemicals. The Chinese are attempting to solve the problem of resistance to insecticide by applying four or five different chemicals in rotation. Although most entomologists are skeptical about the usefulness of such a procedure, the test is admirable, and the long-term results

will be interesting. It is by far the most extensive test of the
chemical rotation technique, in relation to resistance, known to the
delegation.

REFERENCES

1. Burke, J. H., 1967. The Commercial Regions. W. Reuther, H. J.
    Webber, and L. D. Batchelor, Eds., Univ. Calif., Div. Agric. Sci.,
    Berkeley, California. Pages 140-146 in The Citrus Industry, Vol.
    1 (Revised).
2. Woo Xiang-guang, Lai Yu-Sheng and Lin Shan-xiang. 1975. On the
    faunistic changes of insects in the citrus groves of Kwangtung
    Province. Acta Entomol. Sinica *18*(2):209-210.

F. DECIDUOUS FRUITS

Attractive and abundant displays of apples, pears, and other deciduous
fruit in the major markets were an indication of the progress made in
producing deciduous fruit. The orchards visited or observed during our
routine travel were impressive. The varieties grown were attractive,
trees were intensively cultivated, fertility was maintained at a high
level, and the foliage was generally clean and of excellent color. Orchards were often interplanted with other crops, such as vegetables,
legumes, or small grains.

In general, the same insect and mite complexes that are significant
in deciduous fruit culture in the United States are major problems in
China. An exception is the codling moth. It was evident, as in other
fruit-growing areas of the world, that emphasis had recently been placed
on chemical control, with consequent development of insecticide resistance and destruction of natural enemies. As with vegetable insect control, a major effort had been made to reduce the application of
organochlorines, especially DDT. However, DDT is still available and
is used where particularly difficult control problems exist. For example, we observed grapes being sprayed with DDT for control of a large
lepidopterous defoliator. The Concord-type grapes were already changing
color at the time of application. It appeared that, at present, organophosphate insecticides, such as demeton, dichlorvos, parathion, and
dimethoate, serve as the basis for most deciduous-fruit insect-control
programs. On apples, 7 or 8 applications were reported applied for
insect and mite control, in most cases by high-pressure units.

Mites, leafrollers, and aphids appeared to be the most injurious members of the pest complex associated with apples. It was reported that
more than 100 economic insect and mite pests are associated with apple
production, and about 20 species present consistent pest problems. There
were indications that dormant sprays of lime sulphur and soda water are
being used increasingly for control of scale insects, especially San
Jose scale, because these sprays do little harm to scale parasites.

The original fruit moth is reported to be the most serious pest of
peaches. Mites, aphids, scales, leafhoppers, and a pyralid borer cause
much concern. In the Shensi area, generally five sprays are applied for
peach pest control. In the areas visited, it was reported that leafhoppers

and mites are the most persistent problems on grapes and that dimethoate is the insecticide most often used.

The history of insecticide in relation to deciduous fruits in Shensi
Province is so similar to that in the United States and is so indicative
of the urgency for development of integrated pest-management programs
that we are quoting here from a formal paper presented at the Northwest
College of Agriculture, Wukung.

"During prerevolutionary rule without insecticides, diseases and insect pests were severe. There are about 120 species of insect and mite
pests in deciduous fruit orchards of which 20 cause serious damage.
Under Chairman Mao's philosophy an eight-point program was developed
to combat these. During the 1950's the principal apple pests were the
leafrollers *Adoxopyes orana* Fischer von Roslerstams, *Choristoneura
lonicellana* (Walsingham), *Leucoptera scitella* and *Illiberis pruni* Dyar,
and the spider mites *Tetranychus viennensis* Zacher and *Bryobia praetiosa*
Koch. Spraying DDT at budding gave good control after a few years. However, DDT nearly eliminated the natural enemies of many pests including
the predators of the red spider mites, *Stethorus punctillum*, *Coccinella
septempunctata*, *Scolothrips* sp. and parasites of some scale insects.
The red spiders then became the most severe pests. At first these were
controlled with the O-P insecticides such as parathion, but resistance
rapidly appeared and as many as 10 applications were required in a
single growing season. Resistance also appeared to the acaracide tetradifon. The use of insecticides increased until their costs amounted to
25 percent of the production cost of the fruit. *L. scitella* became a
rampant, secondary pest, eating beneath the epidermis and causing severe
leaf drop. After more than two years intensive study of its life cycle,
it was determined that in orchards where broad spectrum insecticides
were not applied, *C. septempunctata* killed 38 to 50 percent of *L.
scitella*. A variety of parasites were also effective, killing up to 25
percent of the overwintering pupae, and up to 74 percent of the fourth
generation. Where broad spectrum insecticides were applied *C. septempunctata* disappeared and the parasitism of *L. scitella* decreased to 1.6
to 6.7 percent.

"On pears, the spraying of parathion (together with DDT) in July and
August was used to control the oriental fruit moth, *Grapholitha molesta*
Busck. These insecticides killed the natural enemies of the San José
scale *Aspidiotus perniciosus* Comstock, causing severe outbreaks where
many trees died. *There are many other examples where insecticides were
used irrationally and caused changes in insect populations* (italics are
ours). Now we no longer use single control measures, but carry out
preventative control measures first together with integrated control.
The present program consists of: prevention by removing fallen leaves,
fruits, and removing loose bark, pruning dead branches; encouraging
natural enemies; trapping moths with black light traps and sugar-vinegar
bait traps, and using insecticides as rationally as possible, avoiding
persistent broad-spectrum compounds.

"On apples the program now involves: orchard sanitation; thinning
flowers and leaves to remove infected ones; picking winter cocoons of
leafrollers; interplanting orchards with legumes to encourage lady
beetles to control apple aphids; examining all new seedlings for

infestations before planting; spraying lime-sulfur during budding; spraying chlordimeform in May, July, and August; spraying trichlorfon where needed for leafrollers; the parasite *Aphytis proclia* may cause up to 80 percent parasitism of San José scale and is encouraged."

G. VEGETABLES

Vegetables are a substantial part of the diet for urban and rural Chinese families. Large vegetable plantings were evident around and within all the cities and towns visited, and individual garden plots were made available to families in the rural areas. These small garden plots are intensively cultivated and the produce represents a substantial contribution to the available source of food for many rural families. Vegetables are being continually harvested and taken to market in the urban areas. Although refrigeration is generally not available, produce in the markets appeared to be of high quality with rapid turnover.

Members of the plant studies delegation said that vegetables are very intensively cultivated. The number of species and varieties was impressive. Persons responsible for vegetable production in the communes we visited consistently reported that 80-110 species of vegetables are grown during the year. In the northern areas, field production was supplemented by greenhouse culture. Vegetables were commonly interplanted on quality soils. Excellent water control and maximum fertilization were practiced. It was pointed out at the Evergreen Szechiching Commune that the matters emphasized in growing vegetables are quality, quantity, diversity of species, seasonal production, and conservation of the soil for future production. The prices received for vegetables had been stable for more than 20 years, with little seasonal fluctuation.

Insects were evident in most vegetable plantings, and controls were generally required to provide a marketable crop. In general, the priority placed on vegetable insect research at the major institutes or at the commune level did not appear to be commensurate with the importance of these crops. However, with the high priority placed on cultural control, a consistent monitoring of insect population, and judicious use of pesticides, vegetable insects and the insect-transmitted plant diseases generally appeared to be under reasonable control.

The major insect and mite problems were associated with the foliage. These problems related both to direct feeding and to the spread of plant pathogens. With the exception of cutworms, soil insects did not appear to be of major significance in any of the areas visited. Spider mites were said to have increased in recent years. There was limited reference to minor, intermittent infestations of root maggots, but no evidence that wireworms or white grubs were important. There was no indication that nematodes were a serious problem in growing vegetables. In fact, there was little indication that nematodes were an economic concern in any of the agricultural areas visited. It seemed reasonable to assume that crop rotation and water management had reduced many of the soil pests to noneconomic status.

Various members of the crucifer family were the most important vegetables. These consisted of various species of cabbage, cauliflower, rape, turnips, broccoli, and so on. The imported cabbage worm, *Pieris rapae* Linnaeus, and the diamondback moth, *Plutella maculipennis* (Curtis), did widespread damage to foliage, and aphids were reported to be a problem, especially in autumn. Aphid-transmitted diseases were reported to be increasing in importance in various cabbage-growing areas. It was pointed out that in the Hunan area, aphids were a serious problem on rape which is grown widely for oil rather than as a vegetable. A generation develops in as little as 5 days in the fall and in 20 days in the summer. Aphids migrated between the rape fields and various weeds. There was reference to leaf beetle injury on crucifers. In the Kwangtung area, it was pointed out that in the past a leaf beetle had been a serious pest, and it was suggested that the increased use of ammonia fertilizer had reduced the infestations of this insect.

Dimethoate appears to be one of the most effective materials for aphid control. Although there was no specific indication that resistance was developing, it was pointed out that there had been a general increase in the concentration of the material applied. DDT had been used in the past for foliage feeders. However, it appears that it had not been used during the past 2 years and had been replaced by trichlorfon, malathion, *Bacillus thuringiensis*, and small amounts of rotenone and phosphamidon.

A large number of species and varieties of cucumbers, melons, and squash were grown in the communes and in individual gardens. Leafhoppers, leaf beetles, and thrips were the most troublesome foliar pests. The thrips problem was of particular concern in the Kwangtung area, where conventional control materials had been unsuccessful and the peasants had developed effective nicotine sprays. These were prepared by dipping tobacco leaves in water to extract the nicotine and then adding the nicotine to lime.

Eggplants were abundant in the field and the market in all areas visited. In the autumn, leafhoppers and planthoppers do serious damage to this crop.

Thrips were reported to be pests of onions. However, the onion maggot does not appear to be of economic importance, and, in most areas, controls are not needed for the general maggot complex.

Aphids, leaf miners and leaf beetles, and leafhoppers are pests of beans. The armyworm and the black cutworm were reported to be pests on most vegetables. These insects have been controlled fairly well by direct application of pesticides, or the use of various baits, including sugar-vinegar bait and bean cake treated with trichlorfon. In the Kwangtung area, we were told about an indigenous insect virus for control of the armyworm. The virus was being evaluated in vegetable gardens near Kwangtung. Since there was a great desire to reduce the use of chemical insecticides, a cooperative research program had been initiated. It involved scientists, third-year students at the university, community leaders, and peasants. Promising results were reported from field experiments in many local brigades, but it was indicated that further study of some of the formulation and related problems would be needed before large-scale commercial use could be proposed.

In summary, it appears that application of insecticides, carefully integrated with cultural control, is the principal method of vegetable insect control. Applications of chemicals were generally made by hand with various types of compressed-air or portable air-blast sprayers. It was evident that insect populations were continually assayed by extension specialists at the production team level, and it was generally assumed that populations were maintained at relatively low levels with this procedure. Persistent pesticides in the vegetable-insect control effort were being phased out, and various insect pathogens were being introduced as substitutes. There was no indication that biological control or host-plant resistance had a major role in the control effort. The Chinese did not appear to be making use of the predator *Phytoseiulus persimilis,* which is widely and effectively used against spider mites in glasshouse vegetable culture in Europe.

H. MAN AND ANIMALS

General

The primary purpose of the delegation's trip to the People's Republic of China was to exchange information on the control of insect pests on agricultural crops. However, we were also interested in obtaining information on insects affecting man and animals, stored-product insects, and structural pests. The tour's chief emphasis was visiting research institutes, experimental farms, and communes; however, the itinerary included two visits to towns to observe and discuss sanitary and hygienic methods for improving human health. Some research on termites and on insects affecting man was reported at the institutes and universities we visited. Observations could be made during the trip on the occurrence of certain insects, such as house flies, mosquitoes, and insects on livestock. Termite research was discussed at two institutes (one in Shanghai and one in Canton). Unfortunately, the trip did not include visits to grain storage facilities, although it was requested. Consequently, no information was obtained for this important area.

Information, mostly limited in detail, was obtained from the following institutions or places:

Peking
Institute of Zoology--Historical mention of DDT and malaria, developmental testing of phoxim.
Szechiching (Evergreen) Commune--Pig, duck, vegetable production, hospital and clinic facilities.
District Market--Sanitation.

Shanghai
Tsaoyang Workers' Village--Sanitation.
Shanghai Institute of Entomology--Termites and mosquito colonies.
Long March (Changchen) Vegetable Commune--Raising pigs.
Nanshang Town--Hygiene.

Canton
Foshan Town--Hygiene.
Institute of Entomology--Termites.
Ssu-hui County--Biological control.

Information was obtained on the species of insects important to these areas by going over insect collections and reviewing available literature.

Research on insects affecting man and animals seemed to be very limited in scope, but it may be that considerable research is being conducted at institutes that we did not visit.

Control of insects affecting man and animals appears to be based mainly on sanitation and source reduction, with some use of insecticides. This approach is in line with the writings and teachings of Chairman Mao and appears to have been unchanged since suggested in 1949. Peasants, farmers, and workers are mobilized to clean up dwellings and streets and dispose of manure and human wastes. At the same time, an educational program is conducted. Charts describe insects and rodents, their life cycles, control and sanitation measures, and therapeutic treatment. Banners urging sanitation and improved health measures are placed in shops, stores, and streets. The system, when coupled with therapeutic treatments available at clinics and hospitals and by "barefoot" doctors, seems to be effective in improving the health and living standards of farmers and other workers. No statistics were obtained on the occurrence of insect-borne diseases.

Insects Affecting Man

Since 1949, the Chinese government has conducted an extensive campaign against public health insects and the diseases they transmit. Basic approaches include: (1) mobilization of peasants and other workers for sanitation and land management programs; (2) educational programs aimed at making the masses aware of the major health problems, the insects involved, and means of coping with these problems; and (3) development of health services for peasant farmers and other workers. This continuing campaign appears to have had impressive results. We, as a group, were shown two cities as examples of progress achieved.

The report on medical entomology in China will deal primarily with this major educational aspect of the program. At the same time, concern over the image produced by this campaign led to frustrations in attempting to understand the status of research and control programs in medical entomology in China. We obtained less definitive information on specific insect problems and diseases than we hoped for. This difficulty may have resulted, in part, from the heavy and hurried schedule of the group and from the fact that primary emphasis was given to visiting agricultural institutes. Most of our questions about insects affecting man were answered only with general statements describing mass mobilization, cleanup, education of the masses, and health services. Consequently, one was left with the impression that more definite information on this aspect of entomology would be difficult to obtain.

Mass Campaigns At all places where questions were asked about insect vectors of human diseases, the same general answer was obtained. That is, before Liberation, sanitation was poor and insects and disease were rampant. After Liberation, use of mass mobilization, education, and health services cleaned up the environment in which people lived and improved health, so that essentially no serious problems remain. Mass mobilization is directed primarily at cleanup of animal and human wastes, proper ditching, good use of water for housing, and destruction of household insects and rodents. Mass education includes use of banners and posters in communes and villages, instructional charts in clinics and hospitals, and education through schools, medical personnel, and mass meetings. Since hospitals are present in communes, and clinics exist at the local level of factories, production brigades, and teams, the combination approach should be highly effective where it is intensively applied. In all the areas we visited, we saw evidence of well-organized health campaigns.

We visited Nanshang Town in Shanghai and Fuoshan Hygiene Unit in Canton to see the hygienic approach being taken in these towns.

Nanshang Town, in the suburbs of Shanghai, contained about 18,000 households, and the population was more than 50,000. We were told that there were 16 factories, 3 primary schools, 3 middle schools, 100 shops, a park, and a hospital. The general introduction to this visit included a comment that diseases and insects were serious problems before Liberation, with little or no hygienic work undertaken. Dirt and filth were in the streets and there was no running water, so that flies and mosquitoes were everywhere. After Liberation the masses were mobilized to do hygienic work with emphasis on eliminating mosquitoes, flies, mice, and bedbugs. Mobilization was accomplished by collective leadership and included cooperative work by town people and farmers. Emphasis was given to prevention by mass cleanup and establishment of health services to instruct the masses.

We toured a cafeteria serving one of the factories, a nursery, a restaurant, a barbershop, a clinic, and a hospital. In the cafeteria, food to be served was kept on screened shelves. The restaurant and barbershop were clean and well organized. In the barbershop, they took pride in showing us that they sterilized the towels in boiling water between use. Both hand and electric clippers were in use. A hospital we visited had 5 specialists employed at the hygiene office to instruct the masses. A clinic we visited had large jars containing mosquitoes, flies, and mice which represent work in collecting and eliminating these pests.

We noted several light traps (most used for surveys), along the routes we followed in the town. They were near canals which would breed mosquitoes. One light trap had live and dead anopheline and *Culex* mosquitoes. Another trap had an electrocutor grid instead of a collecting bag. The clinic listed the following species of insects in their educational material:

Cockroaches: *Blattella germanica* (Linnaeus), *Periplaneta americana* (Linnaeus), *Periplaneta emarginata*.

Mosquitoes: *Culex pipiens pallens* Coq, *Culex tritaeniorhynchus* Giles, *Anopheles hyrcanus sinensis* Wied, *Anopheles hyrcanus lesteri*, *Aedes albopictus* Skuse.

Mosquito control includes sanitation, mass mobilization, use of backpack sprayers to apply dichlorvos to vegetation and trees, and application of trichlorfon as a larvicide. Trichlorfon, compared with other chemicals, is a poor larvicide, and it was surprising to hear that it was used as such.

Night soil was brought to a central cement tank by cart or bucket and allowed to ferment (2-4 days in summer and 2 weeks in winter) before being used as fertilizer.

In Fuoshan City, we visited a hygiene exhibit. This city is 1,000 years old. We were told that before Liberation, conditions were poor and backward, and the industries were destroyed by the Kuomintang. However, conditions improved after Liberation. It is now a city of 230,000 people with 150 factories and heavy industry. The introductory statement on insect diseases and problems was essentially the same as that given at Nanshang Town in Shanghai. However, three processes and seven points were listed as means of improving hygiene and health.

The processes were given as: mobilize the masses to change the environment, educate the masses and get rid of superstition, and carry out provincial policy. The specific points were: get rid of sewage; change to clean environment; treat waste, rubbish, and trash and use them as fertilizer; change dirty ponds to beautiful lakes; install running water; educate the masses; and tear down idols and get used to the scientific habit. The results of this campaign were given as follows: The masses are now accustomed to hygienic methods and like to be clean; the hygienic approach works well; and the masses now realize that Chairman Mao was right.

We visited a part of the town, family residences, and the sewage treatment plant. All were clean and well kept. One of the local hosts rubbed his hand across a cooking top to show how clean it was. Another showed the pipe that had been installed in the cooking area to vent smoke to the outside of the house.

An extensive sewage-treatment plant was visited. Night soil is brought, mostly by boat, and fermented under water in concrete tanks. Methane is collected and is used to run a generator that produces 40 kw per hour. Heat from the generator is used to treat night soil before it is loaded on boats to be used as fertilizer. The sanitation management office here reported that formerly they only collected garbage but have extended their efforts; now they convert waste to useful products. They have 300 workers, 12 large vehicles, and 14 jeeps. Garbage is dumped into large concrete bins. They add 15% water by weight and cover tightly for 4 weeks for fermentation. Temperatures reach about 78° C (172° F). Farmers buy this product as fertilizer.

The treatment of waste was continually mentioned, but the only special treatment observed, other than plain fermentation, was the use of heat.

All the residences visited had bed nets. In response to questions about them, we were told that mosquitoes were not a serious problem but that residents keep them simply out of habit.

Mosquitoes There appears to be little reason to doubt the effectiveness of the mass campaigns to improve health and sanitation. Precise information on the number of cases of insect-borne diseases is, however, not available. Generalizations are made that malaria is still a problem in remote areas and southern parts of China. Filariasis may also be a problem, but information was not obtained. Mosquito-borne encephalitis was mentioned.

An attempt was made to list all the species of mosquitoes mentioned in discussions, listed in educational charts, or observed in insect collections. This information is given in Appendix C.

In northern China, *Culex pipiens pallens* was given as the major polluted-water species; in the South, *Culex pipiens fatigans* occurred in similar situations. Both were mentioned as possible vectors of filariasis.

In Peking, at the Peking Institute of Zoology, use of mosquitoes as test insects for insecticide development was mentioned. A 1973 article in *Acta Entomologica Sinica* describes the synthesis and testing of a series of compounds as insect repellents but it states that the colonies were maintained at another institute.

The only other location visited at which we saw evidence of active research on mosquitoes was the Shanghai Institute of Entomology, Shanghai. Formal presentations did not include information on mosquito research, but during the tour we visited a laboratory maintaining three species of mosquitoes: *Aedes aegypti* (Linnaeus), *Culex pipiens pallens* Cog, and *Anopheles hyrcanus sinensis* Wied. Mosquitoes were used in insecticide evaluations, we were told. They also showed two marker stocks of *C. pipiens pallens*--red eye (recessive) and black head (autosomal recessive)--that they had isolated. They showed us some individuals of this species in which a heterozygous translocation had been induced by low-level radiation. This phase of research was just beginning; they had the translocation only in the males. Although genetic control was not mentioned anywhere on the trip, we conjecture that this institute is starting to experiment with genetic markers and genetic stocks.

Their *Culex* and *Aedes* colonies were fed a diet of wheat flour and yeast. The anopheline colony was fed a diet of liver powder and yeast.

Control of mosquitoes seems to be based primarily on hygiene and water management. Bed nets were seen in residences and hotels wherever we visited in China, indicating, at least, that pest mosquitoes were present and, possibly, that disease transmission is a problem. In the Shanghai area, anopheline and *Culex* larvae were readily dipped from rice fields and canals, but the dipping samples were very limited. Insect repellent--a formulation of dimethyl phthalate--can be bought at drugstores. Dichlorvos for insect control is also available at drugstores. Insecticidal methods of control mentioned during the visit include:

- Use of backpack sprayers to apply dichlorvos to vegetation.
- Application of dichlorvos to straw and subsequent burning of straw.
- Use of trichlorfon as a mosquito larvicide.
- Manufacture of temephos (Abate®) as a mosquito larvicide.
- House spraying of DDT or BHC for malaria control

(principally mentioned as used in the past for malaria control).

Specialized application equipment, such as foggers, truck-mounted ULV equipment, or aerial application equipment, was not observed during the trip.

Cockroaches In southern China, the mobilization campaigns against major pests include mosquitoes, flies, mice, and cockroaches. No special surveys are made for cockroaches; however, they were observed in the train from Changsha to Canton and in the hotel in Canton, including both *Blattella* and *Periplaneta*. In Shanghai, it is possible to buy boric acid tablets at drugstores for cockroach control. Cockroach baits containing dichlorvos were mentioned in discussions.

Other Insects Affecting Man Essentially no information was obtained on the occurrence of other insects or arthropods of medical importance in China. Human lice were mentioned as a former problem. Ticks, chiggers, fleas, sand flies, and others were not discussed in the places we visited.

Livestock Insects

Visits to communes and experiment stations permitted only limited observations on livestock insects. In most areas, particularly in northern China, animals are tended by individuals. Instead of being confined to pastures, animals are staked or tended by individuals and fed on forage along roadsides, banks of ditches, canals, rivers, or levees in rice paddies. In many areas, people are assigned to collect manure from the animals for composting or for direct fertilizing of crops.

Pigs are kept in large concentrations in communes for their meat and for use of their manure as fertilizer. Pigpens visited were cleaned daily and were generally built with concrete floors and concrete drainage troughs. Manure was either collected for composting or held in lagoons for use as crop fertilizer.

The areas we visited, which were well managed, had no serious identifiable problem with flies that breed in filth. The mass-mobilization and education programs for sanitation are highly effective in controlling house flies. Few house flies were observed, although one dairy operation had a moderate number of the flies around animal-feeding areas. An open-air meat market in the middle of the poorer section of a large city had many house flies. The sanitation and education programs appeared effective in the areas where they were applied.

In the one dairy herd visited, a few stable flies were seen on the animals, but no horn flies were seen. Since the occurrence of horn flies would be associated with beef or dairy animals in pastures—and these animals are not pastured in China—it is doubtful that horn flies are a problem where we visited. No direct information was obtained on other insect or arthropod pests of livestock.

House Flies Little more can be added to information on house flies and their control than has already been presented. Outside of China one hears of the control of the house fly in China through sanitation and flyswatters. House flies are controlled to a high degree, but in most places one or two could be easily spotted. Only at a very few places were house flies observed in numbers.

Species of house flies and other related synanthropic flies noted in insect collections were: *Musca domestica vicina, Musca sorbens sorbens, Musca canducens* Walker, *Musca tempestiva, Musca stabulans, Musca hervei* Vill, *Musca amita* Hennig, *Musca bezzii* Patton and Craig, *Fannia canicularis, Fannia priscea, Lucilia cuprina* Weidemann, *Phormia regina, Lucilia porphyrina* Walker, *Lucilia illustris* Meigen, *Chrysoma pinguis* Walker.

I. STORED GRAIN AND STRUCTURES

China produces great amounts of corn, sorghum, and cereal grain, all of which must be stored for varying lengths of time. However, during our trip we were not able to inspect any on-the-farm storage but did see one storage unit off the farm, in Kirin Province. It had several bins of wheat stored in crib-like structures having roofs and walls made of closely woven straw matting. The storage area was exceptionally clean, and there was almost no grain on the ground. This led us to believe that rodents and birds are not serious pests of stored grain in China.

While driving by farms and through towns, we saw many grain storage units. The farm units were either buildings covered with straw matting or circular, bin-shaped brick, or stucco units. In the towns, grain was stored in small silo-type units of brick or stucco inside a fence-enclosed compound.

At the rice-growing commune near Canton, we were informed that they do have problems in stored rice with the confused flour beetle, *Tribolium confusum* Jacquelin DuVal, and with weevils and moths. Farm storage units were said to be carefully cleaned and the walls sprayed with a residual insecticide before new grain is placed in them. The grain then is fumigated with Phostoxin (aluminum phosphide). It was said also that they do not use malathion of dichlorvos strips for protection of the grain, because the above method provides excellent control.

Apparently, tyroglyphid mites are serious pests of stored products in areas such as Szechuan. A recent report deals with the predatory mites of the family Cheyletidae, which are associated with this group of pest mites.[1] Eight cheyletids were reported in Szechuan: *Acaropsis docta* Berlese, *Cheletomorpha lepidopterorum* (Shaw), *Cheyletus eruditus* Schrank, *C. trouessarti* Ouds, *C. malaccensis* Ouds, *Eucheyletia flabellifera* Michael, *E. harpyia* (Rohdendorf), *E. reticulata* Cunliffe.

Among the structural insects, apparently the termites are of the most importance in the South. Research on termites was discussed at the Shanghai Institute of Entomology and the Kwangtung Institute of Entomology in Kwangchow.

The itinerary of the delegation did not specifically include visits to research facilities or problem areas related to these insects. In Shanghai, a request to visit grain storage areas and quarantine operations was denied.

At the Shanghai Institute of Entomology, specialists talked about problems and control. Four genera of termites were first discussed:

Coptotermes, Reticulitermes, Odontotermes, and *Cryptotermes.* Serious damage to houses was mentioned. Chlordane was described as a useful and effective soil treatment. They discussed research on mirex for control of termites and tests involving 100 experimental treatments. At first, they said a 70% mirex powder was dusted on wood. During questions, they also said that mirex was blown into the tubes constructed by termites. Mirex is being tested in the United States as a bait toxicant for termites; blocks of wood that have been impregnated with mirex are put out for termites to attack. Since mirex is slow acting, it is returned to the colonies and slowly kills the colony. But it is hard to see how dusting wood with mirex would be effective. It is possible that termites could become contaminated or receive sufficient dosages to kill colonies. Kills of up to 80% or 95% were reported in Chinese studies.

Some studies on exposing termites to juvenile hormone mimics were mentioned, but no details were given. In the laboratory where termite research was demonstrated, we were shown variations in damage (resistance to attack) in different types of wood. They did not have colonies at the time we were there. They had demonstrations of the life history and forms of the various species of termites, including *Coptotermes formosanus, Globitermes audax, Macrotermes annandalei* (S.), *Macrotermes barynei, Odontotermes formosanus, Reticulitermes flavipes* (Kollar), and *Reticulitermes chinensis* (Snyder).

The other presentation concerning termites was given at the Kwangtung Institute of Entomology. This institute is starting an extensive program on termites. Their work appeared to be largely taxonomic. They had a list of 79 species of termites that occurred in China. (This list is given in the Appendix.) Exhibits showed types of damage, including damage to wire cables, caused by termites. They also had a small colony of termites for observation and testing, and they mentioned the testing of insecticides against laboratory colonies.

REFERENCE

1. Shen Chao-peng. 1975. Preliminary notes on Chinese Cheyletida mites and the life history of *Cheyletus malaccensis* Ouds. (Acarina: Cheyletidae) Acta Entomologica Sinica. 18:316-324.

3

INSECT CONTROL STRATEGIES

A. USE OF INSECTICIDE

General

Modern insecticides have played an important role in the spectacular
improvements in agricultural production and public health that have been
made in China. At the same time, knowledge has been developing about
insect resistance, the rise of secondary pests, the impact of insects
on environmental quality, and ways in which human health is affected by
overreliance on insecticides. By 1975, therefore, China had pioneered
in integrated pest management, conducting widespread programs in which
insecticides were carefully chosen and applied to maximize their action
on the target organisms and to minimize their impact on nontarget species
and on human and animal health. The chief aim is to use insecticides as
significant parts of multifactored pest-control programs to cope with
pest outbreaks. Spraying is generally by spot treatment of heavily infested areas. There is almost no aerial insecticide application except
in some instances on rice and on occasions during outbreaks of the
migratory locust, *Locusta migratoria manilensis* Meyer.
    We found great interest in the development of new and safe insecti-
cides, such as chlorodimeform and phoxim, and in the widespread use of
microbial insecticides, such as *Bacillus thuringiensis* (*B.t.*) and insect
viruses.
    Apparently there was no insecticide production in China before Libera-
tion; everything used was imported. After 1949, production of DDT and
BHC was begun and reached high levels. After the Great Leap Forward
(1958-1959), the organophosphorus insecticide industry developed rapidly.
Presently, dichlorvos, trichlorfon, and dimethoate are each produced in
yearly amounts that exceed the total production of the organochlorines
and of nearly 100 individual insecticides in China. Emphasis has been
on self-reliance; techniques derived from experimentation have been used.
The fast development of the industry is attributed to the socialistic
system with the objectives of "guaranteeing good harvests and diminish-
ing plagues, not for the benefit of certain persons. There are no
patents or secret techniques and production methods are circulated immediately to all the nation." For example, there are presently scores
of factories producing trichlorfon, which appears to be the insecticide

in widest use. We were told that before new products are produced, toxicity is determined and nationwide cooperative field tests are made. An enormous volume of data is secured rapidly. It is considered very important to the future of agriculture that the insecticide industry continue to develop rapidly.

From our visits, seminars, conversations, and observations, the following picture emerges of the present-day use of insecticides:

Organochlorines

DDT is rapidly being phased out of use, largely because of its incompatibility with integrated control programs, its environmental persistence, and insect resistance to it. DDT is restricted to a single application on cotton in most areas for control of the pink bollworm, *Pectinophora gossypiella* (Saunders); cotton bollworm, Heliothis armigera (Hübner); and diamond bollworm, *Heliothis asfulta* Guenee (Figure 3). It was stated that "it is still impossible to avoid DDT completely." Other mentioned uses are on tobacco for the tobacco aphid and the tobacco noctuid, and on grapes, where DDT is used in emulsion with Bordeaux mixture to control leafrollers, and the oriental fruit moth, *Grapholitha molesta* Busck. Use of DDT, as a general national policy, is forbidden on fruits and vegetables, and it is no longer used on rice, where the rice paddy borer, *Tryporyza incertulas* Walker, is highly resistant.

DDT is still used in a limited way for malaria control, and although there are many substitutes, none is considered as persistent as a residual spray for adult mosquito control.

Methoxychlor has never been produced commercially in China.

Benzene hexachloride (BHC) was formerly the most widely used insecticide in China, and 130×10^6 lb were reportedly produced in 1958

FIGURE 3 Spraying cotton with portable hydraulic equipment.

(Cheng, 1963).[3] However, the use of BHC is rapidly being reduced and its use on fruits and vegetables is no longer permitted. Preparations of high *gamma*-isomer content are now favored, and lindane (greater than 99% *gamma* isomer) is available for some uses to diminish environmental contamination by the very persistent but insecticidally inactive *beta* isomer. BHC was formerly widely used as an aircraft spray against the migratory locust, but this insect is now largely controlled by an elaborate pest-management scheme. BHC, 6%, is used with 6% trichlorfon in cinder granules to control the European corn borer *Ostrinia* (*Pyrausta*) *nubilalis* Hübner, and the cotton thrips, *Thrips tabaci* Lindeman. Dusts of 1%-3% BHC are used for control of *Heliothis* spp. on peppers and as a soil insecticide, and 3% BHC plus 1% methyl parathion is used on rice to control the brown-backed planthopper, *Nilaparvata lugens* Ställ; grass leafroller, *Cnaphalocrocis medinalis* Guenee; rice stem borer, *Chilo suppressalis* Walker; rice skipper, *Parnara guttata* Bremer et Grey; and rice thrips, *Thrips oryzae* Williams. BHC is used for control of the pink bollworm in the Shanghai area, both in cotton fields and in storage sheds, but is no longer used on cotton in Kirin Province.

BHC is still used in some areas for control of mosquito and house fly larvae.

Toxaphene is used in a limited way to control bollworms *Heliothis* spp., but because of its high toxicity to fish, its use near water is forbidden.

Aldrin, dieldrin, endrin, heptachlor, and chlordane were never produced on a large scale in China, although relatively small amounts of chlordane are used as a soil poison for termite control. Mirex is used as a dust by mixing the 70% product with "store powder" for blowing into galleries of termites. Endosulfan has apparently not been produced in China. Heptachlor is discussed as a seed treatment for cotton to protect against the wireworm *Agrotis segetum* (Chang *et al.*, 1974).[2]

Organophosphorus Insecticides

These are the basic insecticides in use in China today, and more than 30 O-P compounds are produced industrially. At a major pesticide factory, the Shanghai Agricultural Pesticide Factory, which has a staff of 856, we were told that the 1974 production was 18,148,820 kg. The products in order of largest use were:

1. trichlorfon (Dipterex®)
2. dichlorvos (DDVP)
3. dimethoate (Rogor®)
4. phosmet (Imidan®)
5. fenitrothion (Sumithion®)
6. phosphamidon (Dimecron®)
7. malathion

Trichlorfon is in widest use. Its popularity is due to its safety, lack of persistence, environmental degradability, and low cost. It is described in public banners as "the enemy of a hundred worms." Trichlorfon is the main deterrent against chewing insects in integrated pest management programs on cotton for the pink bollworm *P. gossypiella*, bollworms *Heliothis* spp.; on tobacco for noctuids; on rice for brown-backed planthopper *N. lugens*, rice thrips *T. oryzae* and for other secondary pests. Trichlorfon is used on cabbage and other vegetable crops especially for control of the cabbage worms *Plutella maculipennis* (Curtis) and *Pieris rapae* (Linnaeus). It is also used to control numerous species of leafrollers attacking apples.

The *in vivo* mode of action of trichlorfon is to produce dichlorvos by the following rearrangement (Metcalf *et al.*, 1959):[6]

Trichlorfon is used as a direct generator of dichlorvos, by incorporating it with Na_2CO_3 at 1 to 0.4 and water, for control of the cotton aphid, *Aphis gossypii* Glover. A trichlorfon-alfalfa bait has been used against the mole crickets *Gryllotalpa unispina* and *G. africana* Palisot de Beauvois.

Dichlorvos is widely used as a "residual fumigant" for crop pests, e.g., the soybean pod borer, *Grapholitha glycinvorella* Matsumura. Sorghum stems are soaked in the material and planted at 300 stems per hectare. Dichlorvos is used in cotton fields by infiltration into a porous carrier for control of the pink bollworm after the bolls are closed, and also for bollworms, *Heliothis* spp. On peaches, it is used against the oriental fruit moth. Dichlorvos is used as an aphicide for the cotton aphid, *Aphis gossypii* Glover; tobacco aphid, *Myzus persicae* (Sulzer); cabbage aphid, *Brevicoryne brassicae* (Linnaeus); turnip aphid, *Rhopalosiphum pseudobrassicae* (Davis); and other types of aphids. It is applied as an emulsion to cabbage roots to control the root maggot, *Hylemya platura* (Meigen).

Dichlorvos is widely used as a spray for house flies (for example, Peking District Market is sprayed with it every morning) for local sanitation nearly everywhere in China. In rural villages near rice areas, where mosquito breeding is intense, dichlorvos is sprayed on trees and vegetation as required to achieve control. For indoor use, dichlorvos is applied to filter paper, which is exposed to a draft of air.

The brown longicorn beetle, *Nadezhdiella cantori*, which attacks citrus trees, is controlled by inserting bits of cotton treated with dichlorvos into the holes in the trunk, or water solutions are injected.

Dimethoate is widely used on field and fruit crops for general pest control and seems well suited for integrated pest management. On citrus, dimethoate has replaced methyl parathion, malathion, demeton, DDT, and BHC, use of which is prohibited on bearing trees. Specific uses are for the control of the citrus leaf miner, *Phyllocnistis citrella* Stainton; the leafrollers *Homona coffearia*, *Adoxophyes privitana*, and *Tortrix* spp.; and the mites *Panonychus citri* (McGregor) and *Phyllocoptruta oleivora*

(Ashmead). On deciduous fruits, dimethoate is used for sweet cherry spider mites, *Tetranychus viennensis* Zacher and *Panonychus ulmi* (Koch), and on grapes for *Empoasca* sp. leafhoppers.

On rice, dimethoate is the basic insecticide used in the integrated pest-management program and is especially effective against the green rice leafhopper, *Nephotettix cincticeps* Uhler, which transmits yellow stunt and yellow dwarf viruses, and the rice thrips *Thrips oryzae* Williams. Seedling rice is dipped into dimethoate for systematic protection against attack by the rice paddy borer *Tryporyza incertulas* Walker. On cotton, dimethoate has been the standard control measure for the cotton aphid *A. gossypii* and the spider mite, *Tetranychus bimaculatus* Harvey, but serious resistance problems have occurred.

Phosmet is used on tea at 0.0125%-0.025% to control the scale insect *Lepyrus japonica* (Chou *et al.*, 1974)[4]. Extensive residue and flavor evaluations have been made to ensure its safety for this use. Phosmet degraded by 50% after 1 day and to zero after 9 days. It was not detected in green tea after 7 days, and the taste of the tea was not affected. Phosmet is also used on citrus to control the citrus leaf miner *P. citrella* and the leafroller *Cacoecia asiatica*.

Fenitrothion is considered much safer for general use than methyl parathion, which can be used only by highly trained personnel. Fenitrothion is favored for use on rice, and 3 or 4 applications may be used on each crop; it is especially effective against the rice leafhopper, *N. cincticeps*. When there is heavy infestation by rice leafhopper, brown-backed planthopper, *N. lugens*, and paddy borer, *T. incertulas*, methyl parathion is applied at 1% with 3% BHC as a spot treatment to heavily infested areas of the fields, to avoid serious damage to beneficial insects.

Trithion and malathion are used for citrus scale insects, chaff scale *Parlatoria pergandii* Comstock; red wax scale, *Ceroplastes rubens* Maskell; and cottonycushion scale, *Icerya purchasi* Maskell although these scales are generally under good biological control. Malathion is used against the tobacco noctuid and the tobacco aphid and also for hygiene purposes.

Phosphamidon is widely used on apples, rice, and cotton, largely for control of aphids and mites. It is also used to control the rice stem borer.

Demeton (Systox®) was formerly used on apples and citrus but its use on bearing trees is now restricted. It was also used on cotton to control the cotton aphid. Disulfoton and phorate are used as wheat seed treatments. They are applied at 0.3% of the seed weight to control soil insects and aphids.

Temephos (Abate®) is used to control mosquitoes and has also been used against some phytophagous Lepidoptera. Monocrotophos (Azodrin®) has been used to control bollworms (*Heliothis* spp.) and the cotton aphid.

Fenthion (Baytex®) is used for control of the brown-backed planthopper on rice and apparently has had limited use for control of the tobacco aphid and as a mosquito larvicide.

Among the newer O-P insecticides, phoxim has been developed for many uses, methamidophos (Monitor®) has been proposed for control of the green rice leafhopper (Van *et al.*, 1974),[8] and cyanox has been used for control of the rice stem borer (Van *et al.*, 1975).[9]

Carbamates

Carbaryl (Sevin®) is the only carbamate insecticide in wide use in
China. However, it is not so widely used as the common organophosphorus
insecticides. Carbaryl granules are used to control the European corn
borer, and carbaryl is used as a spray to control bollworms on cotton
(*Heliothis* spp.) and the millet stem borer. An unexpected use was its
employment against the resistant cotton aphid, *A. gossypii*. In response
to a question about injury to the honey bee, *Apis mellifera*, following
the use of carbaryl, we were told that prior to spraying, managers were
warned not to let bees out. An interesting question was raised by the
staff of a pesticide factory as to the purity of Alpha-naphthol, used
in the United States for carbaryl manufacture. They asked whether it
contained appreciable amounts of the carcinogenic Beta-naphthol isomer.

  MTMC (*m*-tolyl N-methylcarbamate) is used to control the rice plant-
skipper, *Parnara guttata* Bremer et Grey; rice leafroller, *C. medinalis*;
and brown-backed planthopper, *N. lugens*.

  Propoxur (Baygon®) is used to a limited extent to control insects
of public health importance.

  The carbamates aldicarb, carbofuran, and methomyl, which are relatively high in mammalian toxicity, have apparently not been considered
for use in China.

Hormone Mimics

These insecticides promote abnormal growth and development in insects,
and their structures generally mimic those of normal insect hormones.
An example is neotenin. China has devoted a large amount of research to
this area, and a substantial number of promising compounds have been
synthesized. The most promising compounds studied are trans-6, 7-epoxy-
1-(*p*-ethylphenoxy)-3,7-dimethyl-2-octene (J-002, J-734), which has been
developed in the United States as R-20458, and trans-7-ethoxy-1-(*p*-ethyl-
phenoxy)-3,7-dimethyl-2-octene (J-738) (*Acta Ent. Sinica, 1974*).[11] Simple
syntheses of these compounds were worked out at the Shanghai Institute
of Organic Chemistry, Shanghai, starting with geraniol, a relatively
inexpensive starting material.

J-734   J-738

  These hormone mimics are routinely sprayed on 5th instar silkworms,
Bombyx mori (Linnaeus), to prolong this instar for an additional day
and thus increase cocoon weights by 10%-20%. However, despite repeated
questioning, we were unable to discover any sign of interest in the use
of these hormone mimics for control of insect pests.

Pyrethroids

We could elicit no information of special interest concerning the new synthetic pyrethroids, although western work in this area was well known. The answer was, "We are looking at them."

Miscellaneous Insecticides

Tetradifion (Tedion[®]) has been used for control of spider mites (*Tetranychus* spp. and *Panonychus* spp.) on apples, but resistance is apparently common. *Chlordimeform* (Chlorophenamidine[®]) is widely used for control of red spider mites on apples and *Heliothis* spp. on cotton, where multiple applications are used. Chlordimeform has also been tried for control of the cotton aphid.

N'-(4-chloro-*o*-tolyl)N,N-dimethyl thiourea (C 9140) has been studied for control of the rice paddy borer and proved to be almost equal to chlordimeform and superior to parathion-BHC mixture.

C 9140

This compound was also effective against the rice stem borer, rice leaf-roller, and brown planthopper. The compound has no contact action but is rather an antifeeding material that affects newly hatched larvae. It is nontoxic to hymenopterous parasites such as *Trichogramma* spp., *Telenomus* spp., and *Tetrastichus* spp. (*Acta Ent. Sinica, 1975*).[11]

Indigenous Insecticides

Large amounts of crude biological preparations from native Chinese plants have been used for crop protection in China. According to Cheng (1963), more than 9,100,000 tonnes of native materials were applied in 1958, or about 27 kg for each acre of cultivated land in China. Plant sources utilized include *Nicotiana*, containing nicotine, used with lime; *Anabasis aphylla*, containing the alkaloid anabasine; and *Derris*, containing rotenone. Garlic was also mentioned, and it was stated that *Kolanchloe* contains extractives very toxic to aphids. This use of indigenous materials has greatly declined with the development of the synthetic insecticide industry. The group saw only one example in Hunan Province where indigenous materials were being applied together with toxaphene, trichlorfon, and dichlorvos to control rice leafrollers. We were told that these indigenous materials are used in southern China, where they are produced in local factories in communes and production brigades.

Basic Insect Toxicology

Apart from toxicological problems discussed elsewhere in this report, basic research on insect biochemistry is conducted at the Peking Institute of Zoology. Some of this work relates to insect cholinesterases as inhibited by various organophosphorus and carbamate insecticides. It was shown (Shieh et al., 1964),[7] that symptoms of house fly poisoning following topical application of dimethoate, trichlorfon, and tri-o-cresyl phosphate are closely correlated with the degree of inhibition of brain acetylcholinesterase but not with aliesterase. In a further study, Leng and Chen (1965)[2] related fly cholinesterase inhibition after poisoning with organophosphorus esters and phenyl N-methycarbamate. Close correlation was found between enzymatic inhibition and symptoms of poisoning. The maximum inhibition of both true- and pseudo-cholinesterases appeared simultaneously and always coincided with the time of paralysis of the poisoned house flies, although the time for maximum inhibition varied with the insecticide used. True cholinesterase showed a higher degree of inhibition than pseudocholinesterase. These studies were highly sophisticated, although more recently emphasis has shifted from this type of basic research to more applied areas, such as the use of juvenile hormone mimics to improve silk production (*Acta Entomol. Sinica*, 1974)[2] and studies of insecticide resistance (*Acta Entomol. Sinica*, 1975).[13]

Repellents for Mosquitoes

Agricultural technology in China provides an optimum environment for the development of biting mosquitoes such as *Anopheles hyrcanus sinensis* Wied, *Aedes albopictus* Skuse, and *Culex pipiens pallens* Coq. These species were prominently on display in working collections in "hygiene centers." We found larvae breeding in ponds, irrigation ditches, and rice fields. Our queries about the current status of malaria in China were generally answered by "It was widespread after Liberation but is now gone," or "Malaria was last present in 1963 and not thereafter." However, in Ssu-hui County, Kwangtung Province, we were told that "the local hospital has reported that they do not have much malaria any more." It is probable that malaria is still endemic in southern China.

Biting by pest mosquitoes still causes much concern. We found hotels in Changchun, Sian, and Changsha equipped with window screens and bed nets but they were still plagued with night-biting mosquitoes. In Sian, we were provided with pyrotechnic "mosquito coils" containing pyrethrins. Biting mosquitoes must be most vexatious in rural China. We saw no window screens in rural areas, but mosquito netting was common.

The repellent demethyl phthalate is produced in China and sold in local stores, although we were told in the Nanshang Village Patriotic Hygiene Unit that repellents were not needed. This was presumably the result of a vigorous village spray campaign in which dichlorvos was applied to trees along the streets.

Research on mosquito repellents has been done by the Section of Insect Repellent Research, Peking Institute of Zoology (*Acta Ent.*

Sinica, 1973).[14] The synthesis of a series of ethyl N-substituted carbamates is reported which were evaluated for repellency to the yellow
fever mosquito, *Aedes aegypti* (Linnaeus), after application to the backs
of guinea pigs. The most active compounds were ethyl N-2-furyl carbamate (K-1023) and ethyl N-phenyl N-formyl carbamate (K-1065).

$$\text{NHCOC}_2\text{H}_5 \qquad\qquad \text{NCOC}_2\text{H}_5$$

K-1023   K-1065

The relationship between octanol/water partition coefficients and log
repellent ratios was roughly linear and favored compounds with lower
partition coefficients. There was no apparent correlation between far
infrared absorption spectra and repellent ratio.

Insect Resistance

The acquired resistance of insects and mites to the use of insecticides
is as troublesome to agricultural production in China as it is elsewhere in the world. The early history of insect resistance in China
has apparently not been documented (Brown and Pal, 1971,[1] Cheng, 1963).[3]
However, we were informed that a colony of BHC-resistant house flies has
been under study in the Peking Institute of Zoology since the 1950's.
Detoxication mechanisms have been studied, but no useful results have
been obtained.

We were told that mosquitoes around insecticide factories virtually
disappeared for many years. But recently they had been observed in
substantial numbers and when evaluated were found to be resistant to
all the products made there, so that multifactored resistance was
clearly present. It was stated that the wide use of insecticides in
agriculture has caused serious resistance in mosquitoes, and this is an
important reason that "we mobilize the masses -- we considered resistance
uses a lot of insecticides, pollutes soil, air, and water -- so we minimize this use."

Alternation of insecticides is rigorously practiced in some areas to
delay appearance of resistance. Apples in Evergeen Commune, Peking, are
sprayed with methyl parathion, demeton, and dimethoate in alternation to
control red apple mites. On citrus in Changsha, Hunan, 4 to 5 kinds of
insecticides are used in rotation against citrus mites and scales.
Determination of resistance and cross-resistance is given high priority.

Cotton aphid Organophosphorus insecticide resistance in the cotton
aphid *Aphis gossypii* Glover is cited as a highly disturbing example.
Before 1958, this insect was controlled by spraying with demeton EC at
0.02%, apparently over a very large area of China (Cheng, 1963).[3]
Resistance developed quickly. By 1962, a dosage of 0.1% was required,
and the aphid showed cross-resistance to dimethoate, disulfoton,

morphothion (Ekatin®), and menazon. However, the aphid was still susceptible to dichlorvos and methyl parathion, which were distributed in areas where resistance was severe. In the Shanghai area, dimethoate is being used for spraying the aphids, but the concentration required has increased from 0.033% to 0.066%, and 4 to 5 applications may be required. Chlordimeform and carbaryl are also used for aphid control on cotton, illustrating the severity of the problem. We were skeptical regarding this use of carbaryl, which is not known for its aphicidal efficacy. Questioning revealed that it was used for its suppressive action at high temperatures.

Laboratory studies in the Shanghai Institute of Entomology show an increase of 5- or 6-fold in the LD_{50} of dimethoate to resistant cotton aphids. Experiments conducted at the Shanghai Institute of Entomology show that omethoate, or O,O-dimethyl S-(methylcarbamoyl)-methyl phosphorothioate, the P-O analogue of dimethoate, is 21 times more toxic to resistant aphids and 9 times more toxic to susceptible aphids than dimethoate (*Acta Sinica Entomol.*, 1975).[13] It was concluded that dimethoate resistance in *A. gossypii* was related to a decrease in the speed of *in vivo* activation, P=S→P=0. Omethoate has been produced in sufficiently large quantities for field evaluation, where its use was shown to overcome the resistance, and it was welcomed by the peasants in the cotton-producing brigades. Commercial production is now beginning. We should emphasize, however, that omethoate (rat oral LD_{50} 50 mg per kg) is about 10 times more toxic to mammals than dimethoate (rat oral LD_{50} 250-500 mg per kg) (Kenega and End, 1975). Therefore, the substitution of omethoate may seriously increase the toxic hazard to workers spraying cotton.

Insecticide resistance has seriously affected other insect control programs, especially on deciduous fruits, where red apple mites are broadly resistant to many pesticides, and in rice production. The rice paddy borer *Tryporyza incertulas* Walker is resistant to both DDT and parathion, and resistance levels of 60 X have been demonstrated to the latter. The rice leafhopper *Nephotettix cincticeps* Uhler is highly resistant to both trichlorfon and BHC.

Basic research on insecticide resistance is conducted by the Research Group on Resistance to Insecticides, Shanghai Institute of Entomology. Laboratory and field studies with the synergist or microsomal oxidase inhibitor O-methylpiperonyl aldoxime, or PA, have been done with dimethoate-resistant cotton aphids, *A. gossypii*, and southern house flies, *Musca domestica vicina* Macquart.

PA

The synergist PA increased the toxicity of dimethoate in the S-house fly 2.6 X and in the R-house fly 1.42 X. Comparable values for omethoate were S-house fly 3.6 X and R-house fly 2.61 X. It is concluded (*Acta Ent. Sinica*, 1975)[13] that the resistance mechanism in the house fly is associated with the increase in degradation of omethoate by the microsomal oxidases. In the cotton aphid, PA decreased the toxicity of dimethoate to 0.1 X in S-aphid and to 0.58 X in R-aphid. PA, however, increased omethoate toxicity 1.3 X in S-aphid and 3.0 X in R-aphid. It is concluded that the resistance in the cotton aphid is the product of changes in the rates of P=S→P=O activation and of degradation of omethoate, both microsomal reactions.

Safety and Selectivity

Agricultural workers in China are greatly concerned about the overall effects of intensive use of insecticides on environmental quality and human health. As a consequence, use of DDT and BHC is rapidly being phased out. In the 1950's, it was decided not to produce the cyclodiene insecticides for large-scale agricultural use because of their lengthy environmental persistence and relatively broad-spectrum non-target toxicity. However, mirex and chlordane are still used for termite control in soil and timbers. There is still substantial concern about these problems, and at the Peking Institute of Zoology, we witnessed experiments in progress with ^{14}C radiolabeled DDT and *gamma*-BHC. Radioautographs were available showing the distribution of these radiolabeled insecticides in carp from frozen sections. There are also GLC studies of the distribution of these compounds in fat, liver, and brain.

With the development of insect resistance to the organochlorine insecticides and concern about their environmental persistence, there was large-scale development of the organophosphorus insecticides, especially methyl parathion and demeton. There is, evidently, much concern about the hazards of these insecticides to users and bystanders. Methyl parathion can be used only by highly trained and experienced personnel. The Ministry of Agriculture and Forestry and the appropriate departments in the various communes have established strict rules pertaining to kinds of application equipment, protective masks, and clothing used. Persons applying dangerous insecticides, defined as having rat oral LD_{50} values less than 10 mg per kg, must carry the antidote 2-pyridine aldoxime methiodide (PAM) with them in the fields. It was admitted that there are occasional cases of organophosphorus poisoning. According to rules at the brigade level, only dilutions, not concentrates, of dangerous insecticides may be stored. All containers must be returned to the cooperative. Thus, there is a high degree of control of dangerous insecticides. A special person is designated to control them in each production brigade.

As a result of these problems methyl parathion and demeton, along with DDT and BHC, cannot be applied to fruit trees bearing fruit. Second-generation organophosphorus insecticides, which have a higher safety margin for the user, are now emphasized for use. Those in widest use are trichlorfon (rat LD_{50} 450-469), dichlorvos (rat oral LD_{50} 25-170), dimethoate (rat oral LD_{50} 250-500), phosmet (rat oral LD_{50} 147-299), phosphamidon (rat oral LD_{50} 15-33), fenitrothion (rat oral LD_{50} 250-670), and malathion (rat oral LD_{50} 885-2,800).

Development of Phoxim

Phoxim, or *O,O*-diethyl phenylglyoxylonitrile oximyl phosphorothionate (rat oral LD_{50} 1,891-2,077), has had very wide-scale evaluation in laboratories and in the field by the Institute of Zoology, Peking.

The story of the development of phoxim is an excellent illustration of China's approach to research and development of new pesticides and is therefore told in detail. Phoxim was selected in 1972 as a safe, broad-spectrum organophosphorus insecticide suitable for large-area treatment of multiple-cropping. In 1973, large amounts were prepared for nationwide testing. Results were very promising, and pilot production, begun in 1974, was welcomed by peasants working in the fields. Large areas were treated by research units of provinces and districts and by experimental units of communes and brigades. Phoxim was evaluated in field tests against more than 50 kinds of insect pests. The rapidity of this development was cited as indicating the efficiency of the socialistic system and of the "3-in-1" coordination of research, production, and application by peasants, cadres, and technical personnel.

The following areas of effectiveness of phoxim were demonstrated in this nationwide screening program with standardized techniques:

Rice--much better than parathion, carbaryl, fenitrothion, and mixtures of methyl parathion and BHC against the yellow rice paddy borer.

Cotton--treatment of 1,320 hectares (20,000 mu) (2,500 hectares during 1974) showed excellent results against bollworms, *Heliothis* spp. It was better than DDT, toxaphene, or dimethoate. It was better than demeton against the cotton aphid.

Tea--effective against Lepidoptera at 1:1500 to 1:2000 dilution and better than trichlorfon and fenitrothion. Phoxim had no effect on tea quality and no adverse effects on silkworms. Residues on tea were found to be rapidly photodegradable, decreasing under natural sunlight from 50.78 ppm (dry weight) to 6.14 ppm after 4 hours and below the limit of detection after 2 days. No phoxim residue could be detected in green tea made after 2 days' exposure. However, under artificial shading, residue decomposition was slower: from 50.78 to 18.57 ppm at 4 hours; to 8.85 ppm at 24 hours; to 4.12 ppm at 48 hours; to 2.14 ppm at 72 hours; to less than 0.50 ppm at 96 hours; and to a trace at 5 days. Therefore, it was suggested that following application of phoxim to tea at 1:800 (50% EC), the safe period before plucking should be 3 days if sunny and 5 or 6 days if cloudy (Insecticide Pesticide Group, Institute of Tea Research, 1975).[10]

Tobacco--phoxim was much better than DDT against budworms, *Heliothis* spp.

Stored grain--phoxim was second only to phenthoate (*O,O*-dimethyl *S*-ethylmercaptophenylacetate phosphorodithioate) against the confused flour beetle, *Tribolium confusum* Jacquelin duVal, and other grain beetles. When it was applied at 5 ppm mixed with grass to grain stored over 1 year, only a few rice weevils were found compared with very heavy infestations in control lots. It was also effective when sprayed on walls of storage bins.

Phoxim was effective as seed dressing for wheat, corn, and cotton, and better than dimethoate. It was evaluated against pests of fruit and forage and against forest insects, mosquitoes, house flies, cockroaches, and bed bugs with good results. Its advantages are: It is a safe insecticide and can replace highly toxic long-lasting residual insecticides in food crops and cotton; its short residual effect permits safe use on tea, mulberry, and vegetables; it does not affect the quality of stored grain, woolens, or soil; and the cost is relatively low.

The residual properties of phoxim have been extensively studied on tea and in potatoes, celery, cabbage, and tomatoes, where no residue could be detected ater 2-5 days. On woolens and stored grain in the absence of light, the residual action is lengthy. When it was applied at 0.02%-0.05% (EC), it protected against the black carpet beetle, *Attagenus megatoma*, and the webbing clothes moth, *Tineola bisselliella*, for more than 40 months. Such applications were reported to be resistant to soap-and-water washing and drycleaning.

The mechanism of phoxim photodegradation has been investigated at the Peking Institute of Zoology, using TLC and detection by serum cholinesterase together with bioassay against *Aeda aegypti* larvae and time of knockdown to *Musca domestica vicina*. After exposure to sunlight and ultraviolet light, four spots showing anticholinesterase activity were detected by TLC. Extraction and infrared analysis showed one to be phoxim. The toxicity to *Aedes* was decreased, but the product was still effective against *Musca*. To the mouse, the toxicity increased from the technical product (mouse oral LD_{50} 1,300-1,400 mg per kg) to the photolyzed product, which produced death in 3 minutes at 396 mg per kg. Irradiation by ultraviolet lamp for 99.5 hours produced a product with a mouse oral LD_{50} of 46.6 mg per kg compared with 725 mg per kg for the nonirradiated product. The toxic symptoms were also different. The photoproduct showed rapid excitation and sudden death; phoxim was nonexcitatory and produced slow death. Technical phoxim gradually increases in mouse toxicity even in subdued light. A 3-year-old sample with an LD_{50} of 1,300-1,400 mg per kg was appreciably more toxic this year. The photodegradation products are yellow, and their greater mammalian toxicity may be an important disadvantage in practical use.

Chlorophoxim, or *0,0*-diethyl 2-chlorophenyl glyoxonitrile oximyl phosphorothionate, has been evaluated to a small extent as a grain protectant on nonfood grain and for treatment of walls. It was less effective than phoxim.

Regulation of Insecticide Use

As suggested above, there is considerable concern in China about the long-term effects of the use of insecticides on human health and the environment. It was stated that the "government pays full attention to protection and control of the environment and the health of the people." Thus, the use of the organochlorine insecticides is being phased out, and integrated control is favored, with emphasis on the development of highly effective insecticides of low toxicity (e.g., phoxim) as replacements and on the use of nonchemical methods when feasible. We were not given specific details regarding federal agencies that control these matters. The understanding of our scientific colleagues in China about them was evidently roughly equivalent to U.S. scientists' understanding of the complexities of our FDA and EPA regulatory processes. The invariable response to our inquiries about regulations and prohibitions was that "these matters are controlled by national authorities," i.e., the Ministry of Agriculture and Forestry or the Ministry of Public Health.

Many entomologists engaged in "popularizing" the use of insecticides (i.e., making recommendations) are concerned about the effects of insecticides on the health of man and animals. Under the developmental process for new insecticides, as described in the discussion of phoxim, residue studies are begun in the small-area tests and extended in the large-area treatments. After popularization, apparently no further checks are made. There is little or no evident feedback from the Ministry of Public Health to these entomologists. The Ministry of Public Health accumulates information as developed in the various Research Institutes about insecticide residues in foods and human and animal tissues and about degradation products. When adverse effects are found, decisions are passed directly to Provincial Revolutionary Committees and then to Districts and Communes, where deleterious practices will be stopped. It was stated that an Institute of Environmental Protection had been very recently established in Peking under the Ministry of Public Health. A system of pesticide residue tolerances is being developed, together with restricted periods between application and harvest, e.g., tea. The Ministry of Public Health also analyzes drinking water for pesticide residues.

Clearly, national decisions regulating pesticide development and use have been made. It was pointed that these decisions are much easier to implement in a socialistic society where the government controls the supply and can stop the use of an insecticide simply by stopping production. Examples of such decisions that were mentioned to us are:

- A decision in the 1950's not to produce the cyclodiene insecticides for agricultural use because of their persistent and toxic residues.
- Control of the use of organic mercurials and arsenicals, which is exercised by the "top level" congressional government.
- The development of highly specific rules for the safe handling of parathion and other toxic organophosphorus insecticides, by the Ministry of Agriculture.
- Prohibition of the use of toxaphene near water because of its high toxicity to fish.
- Prohibition of the use of DDT, BHC, parathion, and demeton on fruits and vegetables.
- A requirement that new insecticides that have adverse health and environmental properties must have national approval for use.
- A policy for deliberate phasing out of the use of DDT and BHC.

REFERENCES

1. Brown, A. W. A. and R. Pal. 1971. "Insecticide resistance in
    arthropods," World Health Organization Monograph Series N. 38,
    Geneva, Switzerland. 491 pp.
2. Chang Gui-Song, Tan Dong-Chang, and Wu Ting-Fang. 1974. Effective-
    ness of cotton seed treatment with heptachlor in the control of
    Agrotis segetum Schiff. Acta Entomol. Sinica *17*:227, 230.
3. Cheng Tien-Hsi. 1963. Insect control in mainland China. Science
    140:269-277.
4. Chou Chen-Hui, Li Wen-Ku, and Chen Tsung-Mou. 1973. Residue
    Studies of imidan on tea bushes, Institute of Entomology, Shanghai,
    and the Institute of Tea Research. Acta Entomol. Sinica *16*:127-132.
5. Leng Hsin-Fu and Chen Dung-Lien. 1965. True- and psuedo-ChE in-
    hibition of house-fly poisoning by organic phosphorus insecticides.
    Institute of Zoology, Peking. Acta Entomol. Sinica *14*:10-14.
6. Metcalf, P. L., T. R. Fukuto, and R. B. March. 1959. Toxic action
    of dipterex and DDVP to the house fly. Jour. Econ. Entomol.
    52:44-49.
7. Shieh Tzeng-Yi, Leng Hsin-Fa, Cheng Dung-Lian, and Chao Yu-Chang.
    1964. Esterase inhibition of house fly poisoning by organophos-
    phorus insecticides. Institute of Zoology, Peking. Acta Entomol.
    Sinica *13*:503-509.
8. Van Teh-Fang, Qui Cheng-Jin, Liou Chen-Kai, and Chu Ke-Ming. 1974.
    Residues of the organophosphorus insecticide "Monitor" in rice
    plants. Acta Entomol. Sinica *17*:409-412.
9. Van Teh-Fang, Qui Cheng-Jin, Liou Chen-Kai, and Chu Ke-Ming. 1975.
    Chekiang University of Agriculture. A preliminary study of the
    Cyanox residue in rice plants. Acta Entomol. Sinica *18*:25-27.
10. Insecticide Residue Research Group. 1975. Institute of Tea Re-
    search. Residue studies of phoxim (Baythion) on the tea bushes.
    Acta Entomol. Sinica *18*:133-140.
11. Institute of Elemental Organic Chemistry, Nankai University and
    Teaching and Research Group of Pesticides, Kwangtung College of
    Agriculture and Forestry. 1975. Experiments on the control of
    rice insects with the new insecticide N'-(4-chloro-*o*-tolyl) N,N-
    dimethyl thiourea. Acta Entomol. Sinica *18*:251-258.
12. Laboratory of Physiology and Pathology, Institute of Sericulture
    Kiangsa; Section of Sericulture Service, Academy of Agricultural
    and Forestrial Sciences, Laboratory of Insecticide Toxicology,
    Institute of Zoology, Peking, and First Research Laboratory Insti-
    tute of Organic Chemistry Shanghai. 1974. Studies on the applica-
    tion of insect hormone analogues to improve silk production in
    Bombyx mori L. Acta Entomol. Sinica *17*:290, 302.
13. Research Group on Resistance to Insecticides, Shanghai Institute of
    Entomology. 1975. Studies on the toxicity of dimethoate and its
    derivatives to resistant and susceptible cotton aphid (*Aphis
    gossypii* Glover) and southern house fly (*Musca vicina* Marg.). Acta
    Entomol. Sinica *18*:259-265.
14. Section of Insect Repellent Research, Institute of Zoology, Peking.
    1973. Chemical structure and biological activity: synthesis of

N-substituted ethyl carbamates and their mosquito repellent activities. Acta Entomol. Sinica *16*:39-46.

15. The Teaching and Research Group of Entomology and Pesticides,
Department of Plant Protection, Kwangtung College of Agriculture
and Forestry. Acta Entomol. Sinica *17*:135-150.

16. Kenega, E. E. and C. S. End, Commercial and Experimental Organic
Insecticides (1974 Revision). Entomological Society America
Special Publication 74-1. October 1970.

B. BIOLOGICAL CONTROL

General

Biological control has received substantial, effective emphasis in China.
Since the Cultural Revolution it has been government policy for scientists to channel their research toward direct, practical assistance to
the rural population and communes. Yet, in this emphasis, the traditional importation of parasites and predators from foreign lands has
received little or no attention. As a part of their entomological program to utilize natural forces and a broad ecological attack of integrated pest control, biological control has been used widely, although
in rather limited ways. It has not been taken as the central feature of
the integrated control approach in the provinces that our group visited.
Rather, cultural control was said by several scientists to form the base
of their integrated control programs. Undoubtedly, more use is being
made of natural enemies than was apparent to the group, because attention is being given to selective pesticide technology. The development
of pest-resistant varieties of crops as a basic activity has had little
support, although lines or varieties of several crop species resistant
to a variety of pests were reported.

It was surprising to find that there appeared to be no highly structured organization of biological control work on a national basis. The
emphasis given to biological control, or to particular types of biological control (manipulation of resident natural enemies--e.g., *Trichogramma*, microbial insecticides, or transfers of coccinellids into cotton
for control of aphids), appeared to be decided rather independently at
the Province and lesser political levels. Within such political areas,
emphasis appears to be dependent, as in the United States, on the interest, competence, and aggressiveness of individual scientists who have
espoused biological control. It would be an error, however, to assume
that national direction is entirely lacking. There is evidence of some
national influence, support and coordination of some areas from Peking.
For example, the government has sponsored the excellent Chinese movie
that is shown in various provinces to popularize the use of *Trichogramma*.

As in the United States and other parts of the world, it is probable
that the scientific expertise needed for implementing biological control does not yet exist in China on a broad geographic basis. Even if

the national leaning were in this direction, the program could be developed widely only as competent, interested, and motivated scientists became available in the provinces to be responsible for the scientific aspects of it. These programs are not transferable to other geographic areas, crops, and pests without considerable experimentation and development by local scientists. In Shanghai, we were told that there is no national agency that determines the advisability of any pest-control measure and then encourages or directs the development of the measure in all provinces. The Chinese do have national conferences at which, or scientific journals in which, such development is reported. If requests are received from an institute or laboratory concerning such development, instructions or published literature may be sent to a province, or a scientist may go there to assist in preparing a program.

The big advantage that China has over most of the world in using specific approaches to biological control is its low-cost manpower. But to utilize its manpower in these programs, each area requires its own corps of scientists.

The vigor of their biological control efforts is seen in several areas. For example, it is seen in the widely developed and reportedly effective mass releases of *Trichogramma*, in the more restricted releases of *Anastatus*, and in the distribution of strains of *Bacillus thuringiensis* (somewhat widely) and of *Beauveria* (more restricted). These methods have been used against a number of major pests. They also use ducks for insect and weed control in rice. Other reported efforts that were not seen include, for example, release of the "red beetle," *Rodolia*, for control of scales on citrus in Hunan Province. The successful use of *Trichogramma* and *Bacillus thuringiensis* has been a real stimulus to other biological control efforts. Biological control is an important part of integrated control in China, along with, but secondary to, cultural control.

This report is restricted to the efforts that our group was able to observe or discuss with Chinese scientists. Many other efforts in biological control are undoubtedly under way. This account should be taken only as an indication of the scope and depth of Chinese efforts in this area.

It is perhaps significant that with respect to biological control approaches, Chinese scientists had a close association with the scientific progress of the U.S.S.R. during the 1950's. The strong emphasis that we now see in China on the use of *Trichogramma* spp. and *Beauveria bassiana* (Bals.) V., in particular, may stem from that association (as well as from internal political emphasis suggesting the use of natural forces). The emphasis on *Bacillus thuringiensis* must be derived at least in part from the work of E. A. Steinhaus in California and consequent development of this bacterium for use in several countries, especially the United States. The extensive research on and use of *Trichogramma* in the U.S.S.R. is a matter of record (e.g., Klassen 1973)[9]. Yet China has also made innovative and solid accomplishments on her own in this area (Collaborative Research Group on Biological Control of Rice Pests, Kwangtung Province 1974)[2].

Trichogramma

Among some 12 species of *Trichogramma* reported in China (Pang and Chen, 1974), four (*T. dendrolimi*, *T. australicum*, *T. ostrineae*, and *T. japonicum*) are mass-produced in special institute laboratories and in various communes. They are released extensively for control of such varied pests as the European corn borer, pine caterpillars, rice leafrollers, sugarcane borer, and other Lepidoptera in cotton. The host insect eggs used are those of the giant silkworm moths, *Samia* (=*Philosamia*) *cynthia* (or *S. ricini*) and *Antheraea perniyi* and the rice grain moth *Corcyra cephalonica*. In most of the world *Sitotroga cerealella* is the host used in most programs. It was reported in Peking that *Trichogramma* has been released in some 663,000 hectares of cotton yearly. *Trichogramma* are released in rice at about 300,000 ca per .39 hectare (50,000 per mu) at 30 points per .39 hectare and five times for a single rice crop if densities of target insects (e.g., rice leafroller, *C. medinalis*) are high. Lower numbers are used to control low density infestations. The Chinese commonly get about 80% parasitization or better, and this is adequate, with costs about half that of chemical control, we were told. On the other hand, for the European corn borer, *Trichogramma* are released only three times at about 130,000 per .39 hectare with parasitization averaging 75%. In a one-generation area, the cost is about $1.2 yuan per .39 hectare (20 cents/Chinese mu), not counting the labor of distribution. This latter costs about $0.52 (1 yuan) to cover about 3.12 hectares (50 mu).

Trichogramma Use in Kirin Province

For the European Corn Borer We did not see *Trichogramma* being used for control of European corn borer, but releases were reported to be made in corn of about 90,000 to 120,000 *Trichogramma* per .39 hectare, and the distribution is at 30 points per .39 hectare (5 mu). They start releases before moth egg laying begins and continue through the period of peak moth egg laying. The reader is also referred to two papers on this subject: 1) Seed and Plant Protection Station, Tung Hua Region, Kirin Province (1975)[17], and 2) Agricultural Experimental Station, Sanke-Yushu People's Commune, Tung Hua County, Kirin Province (1975)[1]. *Trichogramma* are used in parts of Kirin in combination with *Bacillus thuringiensis* at different times during the season. In single brood areas *Trichogramma* are usually sufficiently effective but *Beauveria* may be used as a supplement. Cost of the *Trichogramma* program is about 1 yuan per .39 hectares.

For production of *Trichogramma* in Kirin Province, the county produces the host eggs. These are presumably eggs of the giant silkworm moths, *S. cynthia* or *A. perniyi*, which are used commercially for silk production, and the supply of eggs is thus largely a by-product of the silk industry and very cheap. The commune centers parasitize the eggs used in fields and send them to the brigades which make the releases. Results of one evaluation show that corn borer populations per 100 plants were reduced an average of 76% for 4 years (1971-1974), compared with controls. The total acreage under the program in three Kirin counties was increased from 71 hectares in 1971 to over 34,710 hectares in 1974.

For the Spiny Pine Caterpillar The Kirin Institute of Forest Research was built in 1956. Since 1964, it has devoted much effort to use of *Trichogramma* control of the spiny pine caterpillar, *Dendrolimus sibericus*, a pest of all pine species in the province and of larch and spruce. Former non-biological control methods for control of *D. sibericus* were not satisfactory. They have now developed a satisfactory program that includes some chemical control but to an increasing extent uses *Trichogramma*. It was here that methods of mass-producing *T. dendrolimi* on eggs of the giant silkworm moths were first developed.

Two forms of *Bacillus thuringiensis* were tried in Kirin forests with good results (kills of 94% and 91%). But their use is not considered practicable because much of the terrain is very rough and inaccessible for spray rigs. Airplane treatments do not give adequate coverage in Kirin's dense forest plantings. (The successful results reported from aerial spraying of *B. t.* in Hunan forests can be explained by the much sparser tree growth in Hunan.) Hence, in Kirin they concentrated on use of *Trichogramma*, which can be distributed easily. The *Trichogramma* disperse themselves to some extent after "delivery." Entomologists are trying to avoid use of chemicals, believing that chemicals bring on various other insect problems and resurgence of target pest populations as a result of reduction of natural enemies in the forest.

Many hosts for *Trichogramma* were tried, but entomologists settled on the giant silkworms. They are good hosts for mass production, the eggs are readily available the year round, and one egg produces an average of about 60 *Trichogramma*--as many as 175 have been recorded. They produce large, healthy parasites that can lay many eggs. If the ratio of parasitizing *Trichogramma* to host eggs is kept at a proper level (1:2) in the oviposition chamber, there will be few eggs receiving a single or very few parasites (otherwise the adults may emerge as "bloated, inactive hulks"). The fecundity of *T. dendrolimi* is 105 as an average.

At the production facility, there is an automated belt equipped with flanges and flaps for holding a large supply of giant silkworm coccoons that are nearing the time of emergence (Figure 4). The belt can be rotated by a motor to put any of the freshly emerged moths into position for collection. Virgin females are collected and passed through a good sized electric meat grinder. This squeezes out eggs and all body parts into a container. Some eggs are broken, but most come through intact. Since they are infertile, they can be stored and used over a period of time for production of *Trichogramma*.

The eggs are separated from the body "gunk" and cleaned by washing and screening processes that we did not see. The eggs are dried and stuck to paper sheets or cards (Figure 5) for short-term storage or for immediate parasitization to maintain the *Trichogramma* cultures (Figure 6). The main supply may be sent to parasitization units at the county or commune levels for *Trichogramma* field production or forest release (in this case for forest release). The host eggs do not remain suitable for parasitization after long storage. Where long storage is

94

FIGURE 4 Automatic
production of silk-
worm moths for egg
production.

desired, host pupae, rather than eggs, are stored. For this they used
a large glass-enclosed chamber for the *Trichogramma* oviposition unit
into which egg cards are placed.

Some *T. dendrolimi* are reared in an outdoor screened insectary under
natural variable temperatures and humidity. This is to expose the parasite to more normal conditions for feeding back "hardier" individuals
into the main stock culture. Entomologists claim a 10% per year input
to the present stock in this way. This, however, probably would not
remedy the main problem that might arise from continuing isolated culture of such an insect in the laboratory.

The period for releasing *Trichogramma* is very important. Emergence
from the egg cards must coincide with the ovipositioning period of the
pine moth. They use about four release points per .39 hectare (10 per
hectare) and an average of 1,000 *Trichogramma* per point, or about 4,000

FIGURE 5 Silkworm
eggs prepared for
parasitization to
maintain *Trichogramma*
cultures.

FIGURE 6 Production of *Beauveria* at the commune level.

per .39 hectare. In experimental releases they commonly get 95% control, but for large-scale forest management releases they get about 80%. They also vary the release rate (number of parasites and number of times released) according to the density of the caterpillar, the level of natural parasitization by a species of *Anastatus*, and the sex ratio of the *Trichogramma* being released.

Natural mortality of the spiny pine caterpillar varies from a maximum of 40% (mainly from parasitization by a species of *Anastatus*) to a negligible level. The cost of production for pine caterpillar control by *Trichogramma* is .56 yuan per .39 hectare. This is .68 yuan per .39 hectare, less than the cost for corn borer control because fewer numbers are needed per release and fewer releases are required per host generation. The Chinese showed us an adjacent demonstration stand of young larch (a good host of the pine caterpillar) where they started releasing *T. dendrolimi* in 1965. The density of the caterpillars then averaged 40 per tree, and 60% of the trees were infested. In 1975 the density had declined to 2.6 per tree and 6.2% of the trees infested. They considered that this control has been largely the result of the *Trichogramma* releases. They did not need to, and did not make any releases 1975.

Although the production method using oak silkworm eggs for *Trichogramma*, and its utility, were developed in this forestry unit, many areas of agriculture in Kirin Province became interested in the method. It was the Kirin Academy of Agricultural Sciences that "popularized" the use of *Trichogramma* in the province, including use against the rice leafroller since 1971. Since the method has been popularized and shown to be cheap, deliverable and effective, our hosts implied that they probably will not likely try to mass produce and release the endemic, somewhat naturally effective parasite, *Anastatus* sp., of the spiny pine caterpillar.

Trichogramma Use in Hunan, Tangwan and Kwangtung Provinces

In the Shanghai area, Trichogramma are used against certain pests of rice, cotton, and some truck crops, perhaps to a considerable extent. The Agricultural Exposition there has a display illustrating their use against these pests and the Entomology Institute mentioned their use against "several" pests. Nothing was said about the use of Trichogramma at the Long March People's Commune. The Tangwan People's Commune in Shanghai said that they use Trichogramma against the rice leafroller when it is necessary, but usually it is not a bad pest. The "middle-school" children there produce the Trichogramma. More experienced technicians handle the production in Kirin Province.

In Hunan, a main rice area of China, Trichogramma are used primarily for control of the rice leafroller Cnaphalocrocis medinalis Guenee on about 6,435 hectares (100,000 mu) of rice. Both T. australicum (the most effective species) and T. dendrolimi are used. The methods of production developed in Kirin are used (i.e., use of giant silkworm moth eggs as hosts, etc.). The rate of distribution at 30 points per .39 hectare depends on the density of leafroller eggs, and the parasitism achieved often ranged from 70 to 85 percent.

If leafroller eggs are less than ca 300,000 per .39 hectare (50,000 per mu) they commonly release about 60,000 Trichogramma per .39 hectare (10,000 per mu).

If leafroller eggs are ca 300,000-600,000 per .39 hectare they release ca 130,000 Trichogramma per .39 hectare.

If leafroller eggs are greater than 600,000 per .39 hectare (100,000 per mu) they release 180,000-240,000 Trichogramma per .39 hectare.

In Hunan, Trichogramma are also released for control of, and are effective against, a sugarcane borer, Argyroploce shistaceana Snellen, European corn borer and pine caterpillar.

Trichogramma are also widely used in Honan Province which we did not visit. In 1974, they were used in 2,925 hectares in Lai-kan County alone against the European corn borer in corn, with good results. This work is reported under the Grain Insects section of this report.

In Kwangtung Province (Changshan University) studies looking toward use of Trichogramma were conducted in the early 1950's, and by 1956 this parasite was widely used against the sugarcane borer. No data on per-cent kill was given. In the early 1950's, the Institute of Insect Ecology of the Department of Biology of Chungshan University worked extensively on methods to break the diapause of the giant silkworms Samia cynthia and Antheraea perniyi enabling them to rear 2 generations in a year. They tried (1) rearing from the third instar under a photo-phase of 19-21 hours and, (2) storing the pupae for one month at 10° C. The latter method was best. The eggs of these moths are widely adapted for rearing, not only of Trichogramma, but also of Anastatus sp., used for control of litchi stinkbug Tessaratoma papillosa.

The university later developed a distinct biological control labora-tory which in 1972 became the Kwangtung Entomological Institute. This institute does extensive work on Trichogramma for use against rice pests and sugarcane borers, and is conducting a considerable range of other studies. It was here that the excellent Chinese movie on Trichogramma

was produced. It is highly significant that *Trichogramma* is reported
to be used yearly on about one-fifth of the rice acreage in Kwangtung
Province, and in two-thirds of the counties (196 stations). It is the
accepted method of combating the rice leafroller. In this institute,
4 species of *Trichogramma* are studied, and their culture methods are
being developed for commune use. These are *T. australicum*, *T. ostrineae*
and *T. dendrolimi*, reared on giant silkworm moth eggs (above), and
T. japonicum which is more host specific and is reared in the laboratory on eggs of the rice grain moth, *C. cephalonica*. *T. japonicum* is
considered to be the most effective, although it cannot be multiplied
as rapidly and its use has not been so widely popularized. Use of
Trichogramma for rice pest control and for the sugarcane borer is considered a major success.

The successful use of *Trichogramma* spp. is reported to rest on (1)
suitable production and liberation methods, (2) adequate short-term
prognosis of weather conditions and target insect status, and (3) use
of the most effective species, and/or strain for the area and crop.
Studies were made on the temperature and humidity relationships of the
Trichogramma species considered useful in this Province. Against the
rice leafroller in general *T. japonicum* is efficient mainly south of
the Yangtze River, *T. australicum* south of the Yellow River, and *T.
dendrolimi* throughout China where this pest occurs.

Temperature (°C) Relations of Three Species of *Trichogramma* Used in
China

	Min. dev. temp.	Opt. dev. temp.	Min. comple- tion temp.	Day°	? Survival %
T. japonicum	11.09	27	15	134.4	+ 80
T. australicum	11.01	23	15	155.85	80-90
T. dendrolimi	10.34	25	10	161.36	80-90

In general, good stable control was reported by use of *Trichogramma*
--as good as or better than control by use of chemicals. In 1971-1973,
its use in the Lang Tung Commune was extended from limited use to over
1,287 hectares, or over 70% of the total area in rice. Parasitism in
most of the paddies for those three years exceeded 80%. It was reported that the leaf damage in 1971 in the absence of *Trichogramma* releases was commonly 20-25% and that with the yield loss was close to
20%. Experiments in 1970 and 1971 showed that leaf damage could be
reduced from more than 20% to about 1% by releasing *Trichogramma*. The
releases have cut insecticide use by half, a fact reported to be much
welcomed by the paddy workers, whose personal hazard is thereby reduced.

Studies of the effect that *Trichogramma* releases against the rice
leafroller has on other rice insects are being conducted. In one such
area where releases were properly made from May 5 to June 3, 1974, and
in another area where releases were properly made from May 18 to May 30,

1974, parasitization of the rice green caterpillar, *Naranga aenescens* Moore, was 91.7% on June 19 in the first area and 83.9% on June 22 in the second. In other areas where the brigades had released very few *Trichogramma* (there were no areas where none was released), parasitization of *N. aenescens* was only 10%-15%.

Trichogramma releases have recently formed a significant part of experiments to test the effectiveness of an integrated approach to control of the whole complex of rice insect pests in Kwangtung. They made releases for control of the rice leafroller, rice green caterpillar, and yellow rice paddy borer. Some 30,000,000 *Trichogramma* were released in these experiments, involving 108 hectares (1,605 mu) of rice. While the chemicals used in this experiment for control of some of the other pests would be expected to depress the potential effects of *Trichogramma* to some extent, damage by the yellow rice paddy borer was reduced by the various measures from about 4% to 1% and less. Parasitism by *Trichogramma* of both the rice leafroller and the rice green caterpillar was about 70% in 1974. In 1975, however, heavy rains reduced the parasitism considerably.

In Tahsia (Big Sand) Commune in Kwangtung Province, we were shown the commune's production of *Trichogramma*. Their production of *T. japonicum* on the rice grain moth, *C. cephalonica* was especially interesting. This species is said to be the best parasite for use against rice pests in this area, although *T. dendrolimi* and *T. australicum* are more commonly used in Kwangtung and other provinces. The rice moths are reared in a rice meal medium placed in reed trays to a depth of 1 to 1-1/2 inches. The moths are then housed in a darkened chamber kept at 26°-30° C (79°-87° F) and 75%-80% relative humidity. The eggs of the rice moth are roughish on the surface and larger than those of *Sitotroga*, which are commonly used in the U.S.S.R. and in North America for *Trichogramma* rearing. They are, however, much smaller than the smooth eggs of tha oak silkworm moths and produce fewer *Trichogramma*, a fact that accounts for the lower mass-production potential said to apply to *T. japonicum*. When the moths emerge, they are collected and placed in screen sleeve-cages where they lay their eggs on the screen, from which the eggs are brushed daily, with a clothesbrush or hairbrush, onto a paper. Later they are cleaned of debris and sprinkled onto and stuck to glue-coated paper sheets 10 inches square in size. One sheet contains enough eggs to produce 60,000 *T. japonicum* (about 2 to 5 per egg, compared with an average of 50 or 60 for an egg of the oak silkworm moth, where *T. dendrolimi* are used). The sheet of parasitized eggs is cut into squares of 5,000 *Trichogramma* each for field release. From 120,000 to 300,000 per .39 hectare are used, the number depending on the species of *Trichogramma* and the density of the leafroller. From 180,000 to 240,000 per .39 hectare are not required of *T. japonicum* as appears to be so from *T. dendrolimi* and *T. australicum* when leafroller egg densities are over 600,000 per .39 hectare. An excellent bulletin with color plates is available for instruction of commune workers and specialists. The commune still uses *T. dendrolimi* and *T. australicum* in most of their acreage and they are reared as described for the Kirin Province production. *T. japonicum* is now ready to be "popularized."

Microbial Control and Insect Pathology

Basic studies are being conducted on the discovery, characterization and pathology of new strains or species of bacteria, fungi, and viruses of potential use in insect control in China. However, all this work is closely oriented only toward use in microbial control programs, and the basic studies are therefore reported under the above heading.

At the Peking Institute of Zoology, a brief account of some of this basic work and its utility in practical programs was presented.

Basic studies are conducted on entomogenous bacteria (mainly forms of *Bacillus thuringiensis*--*B.t.*) and viruses (NPV attacking armyworms and cotton leafworm, *Prodenia litura* (Fabricius), and a granulosis virus of cutworms, e.g., *Agrotis* spp.).

At the Zoology Institute in Peking, basic studies on *B.t.* have identified, and in part characterized, 17 types of *B.t.*, involving 12 distinct serotypes. Materials from different geographic areas are screened and studied. Scientists have studied 53 "strains" in the past 2 years and found them to vary considerably in virulence for various insects. The house fly is susceptible to *B.t.* if treated in the third instar, but not in the adult stage. Adults that emerge from such treated larvae are deformed and do not reproduce.

The Chinese discovered and characterized a new, promising form or variety, *Bacillus thuringiensis ostrineae*. (This may be the form being produced in Tahsia (Big Sand) Commune.) Methods for producing *B.t.* were studied as to their effects on beta-exotoxin production of various *B.t.* types. Of 40 cultures studied, only 5 were found to develop beta exotoxin. For quality production, it is necessary to have proper culture conditions, including temperature, moisture, and time of fermentation. The Chinese have studied "semisolid" and liquid media for production. The liquid fermentation gave a product that increased insect mortality. But it is considered too expensive for general use on the farms, whereas the semisolid production is cheap, costing about half as much as conventional insecticides. The most effective culture time for peak rate of production of exotoxin was 24 hours. We learned later that the Tahsia (Big Sand) Commune in Kwangtung Province uses a semisolid culture time of 30 hours. The effectiveness (mortality) was considered to be as good as that of the liquid fermentation material; both materials were produced by the commune.

B.t. formulations are widely used as selective insecticides in a number of provinces, e.g., in Shansi. They are widely used in Hunan and Kwangtung Provinces against such pests as the pine caterpillar, *Dendrolimus punctatus* (90%-100% kill), diamondback moth, and cabbage armyworm (80%-100% kill). Aerial spraying was effective in a test on the two last-named pests. *B.t.* has also been used for control of rice stem borers, leafrollers in rice, and rice skippers. In the past 4 years, 1,100 metric tons of such microbial materials have been produced and used in the rural areas, e.g., on about 12,800 hectares of cotton. It was reported that various *B.t.* strains are widely used in integrated control programs and are more effective against some pests if used in combination with low-dosage applications of conventional insecticides, e.g., for the European corn borer. Mortality of 60%-80% is considered adequate in some situations. In Kirin Province, *B.t.* is used for corn borer control in the early stage of crop development and *Trichogramma* later in the season. Moreover, during our visit there, *Beauveria* was being used effectively for corn borer control in early August in the late-season stage.

At the Peking Institute of Zoology, studies and uses of viruses in China were also reported. Twelve types of viruses have been isolated and studied in China, and extensive electronmicroscopy work on them is being done. Their electron microscope was made in China, and it gave good magnification and resolution. Basic studies were being conducted on virus breakdown under exposure to light. NPV virus put on the undersides of leaves was effective against the cotton leafworm, *P. litura*, for 7 days, but effectiveness then declined rapidly. When put on the upper sides of the leaves, effectiveness declined rapidly the first day. Under ultraviolet lighting, effectiveness declined 90% in 10 minutes and completely in 30 minutes (Hwang and Ding, 1975)[8]. In answer to a question, our hosts said that they have not yet begun studies but are contemplating incorporation of light-barrier materials in virus-spray formulations.

Bioassays have revealed results of 40%-100% mortality with use of a nuclear-polyhedrosis virus of the cotton leafworm, and more commonly results of 75%-98% mortality when used on cotton in the Kwangchow area. This virus has no effect on *A. ypsilon*, *E. segetum*, *M. separata*, and *B. mori* (Hwang and Ding, 1975)[8].

In Changchun, Kirin, at the Institute of Plant Protection of the Kirin Academy of Agricultural Sciences, work on biological control using *Beauveria* and *B.t.* was mentioned. But we were unable to see it being used in combination with *Trichogramma* for control of the European corn borer.

In 1961, there were only four units in Kirin working on biological control--three on *Beauveria* and one on *Trichogramma*. Now there are 200 units--120 on *Beauveria* and 80 on *Trichogramma*. About 400 people are working in these two areas. The *Beauveria* work is reported here.

Use of *Beauveria*

Beauveria bassiana was isolated from an endemic infection of the European corn borer in Kirin in 1970, and its utility was quickly investigated (Hsiu *et al.*, 1973)[5]. A granular powder preparation is used in the fields. We saw what was apparently excellent control of the European corn borer by use of *Beauveria* on the experiment station grounds. The material had apparently been applied manually to the silk of each ear of corn. Studies are under way to see if *Beauveria* (or *B.t.* or *Trichogramma*) may be used against soybean insects instead of the cultural and chemical controls now mainly used.

The European corn borer is a pest of many crops, but mainly corn, millet, and sorghum in Kirin. Before the Cultural Revolution, it was controlled mainly with chemicals, but with the emphasis since then on integrated control and use of natural forces, biological control has been developed and popularized. The use of *Beauveria* and *Trichogramma* combined was expanded from about 195 hectares in 1971 to about 25,740 hectares in 1974 and to about 57,720 hectares (900,000 mu) in 1975.

The field efficiency of 5%-10% granular preparations of *Beauveria* against the corn borer was reported to be 80%-90%, with an average of 82%. The fungus can propagate itself in the field under Kirin conditions. An extended side benefit is thus realized from a single treatment; it affects the next host generation.

There are two principal methods of artificial distribution or treatment with *Beauveria*. In the fall, workers gather and pile cornstalks, stubble, and residue, then spray these piles with *Beauveria* to kill the overwintering larvae in the residue. The efficiency of this operation is 83%-96%. It was reported that by this method the survival of a fall population in an area is reduced by 75%. Apparently all the stubble is not so treated, or other host plants not treated in the area may account for some survival. The success of this effort is said to be in its use in integrated control.

Integrated control is an objective. Some chemicals are used, especially for the second generation of the corn borer, and pheromone trapping studies are being developed. It was reported that the pheromone trapping studies are not far enough along to determine the utility of the traps for monitoring or direct control, and thus to help in guiding the use of *Beauveria*, *B.t.*, *Trichogramma*, or chemicals, or to serve as a supplemental direct-control measure. It was reported that *B.t.* and *Trichogramma* are also used for corn borer control in parts of Kirin. Combined use of *Beauveria* and *B.t.* eliminated all corn borers in one test (Hsiu *et al.*, 1973)[5]. Moreover, the stripcropping and the rotation schemes used widely in Kirin could serve as other components of an integrated program, and so could any resistance their corn lines may exhibit.

If the corn borer threat is not severe, and there is adequate rain, they use *Beauveria* only, but if populations are high, they use carbaryl plus *Beauveria*. The synergistic effect is said to be about 10%. The synergistic effect of *Beauveria* with BHC was also studied and was about 20% (Hsiu *et al.*, 1973)[5].

Beauveria Production

Beauveria is produced in the countryside by the commune pest-control specialists, technicians, and workers, with guidance or collaboration furnished by the Institute of Plant Protection in Kungchuling and the country production specialists (Figure 6). The basic culture is produced at the county level, then increased to the amounts needed by a commune at the commune level. The distribution and some of the operations are done at the brigade level.

At the outset of this work in Kirin, they sought the simplest practical method of production, using raw materials produced on the farms. They have used variously, apparently, wheat bran, rice powder, cornstalk powder (or material of other grasses), and compost humus. Cultures may be developed in large, flat trays or in glass flasks and jars in a culture room or on the ground in shallow, sheltered pits outdoors. The final granular or powdered product is stored in plastic bags.

The Ta Yushu Production Brigade of Nan Hwei-tzu Commune near Kungchuling, Kirin, expands its own *Beauveria* for field use against corn borers. One method uses a mixture of wheat bran and rice bran (80:20) in flasks and jars. The workshop uses about seven staff members in December and January and again from March through July. They produce about five kilos of "second generation" *Beauveria* a day. This is increased by 10:1 for 50 kilos a day of "third generation" material for field application. The yearly production of the final generation is about 1.5 metric tons.

A second method was to mix the inoculum with the culture medium and place the mixture in shallow ground pits outside. The whole is covered with straw mats and then dirt, and with a sheltering roof of mats for the culture period. The more conventional production is in culture rooms in crocks.

This commune began using *Beauveria* in 1973, when corn borer damage was sometimes reported at 50%-60%. It now causes no more than 15%-18% damage (presumably infested stalks--not yield loss). A field density sample in 1973 showed 30 per 100 plants, but a sample in 1975 showed only 5 to 7 or fewer per 100 plants.

For field use, the preparation is commonly mixed 1:10 with fine sand for dusting at 2 g per plant on the forming tassels or silks of corn about 7 days before open flowering. Since an insecticide is applied in a granular or similar form, the costs are about the same. Aerial applications of *Beauveria* are made in one province, but the efficiency is less (60%-70% mortality).

Use of *Beauveria* presents four advantages: efficiency, no hazard to workers or the environment (our EPA might think differently), simplicity of production (can be done by brigade workers), and low cost. They were conducting an experiment in 1975 in which half of an area was being treated with chemicals and the other half with *Beauveria*. The results are not yet available.

In Kwangtung Province, *Beauveria* is produced in the communes by both the laboratory and "pit" methods, as in Kirin, to obtain material for control of the corn borer. But in Kwangtung, *Beauveria* is used to control a pine moth, *Dendrolimus punctatus*. In Kirin, they use *Trichogramma*, rather than *Beauveria*, to control the related species, *D. sibericus*.

Miscellaneous Uses of Pathogens

In Shanghai, scientists have a problem with the tussock moth on mulberry that also attacks a few other trees. Its urticaceous hairs affect many people, and a highly selective control method is needed --one that will not affect silkworm culture. A microbial control method is being developed; a nuclear-polyhedral virus (NPV) is applied at 15,000 particles per milliliter. The material is applied at the third instar stage, and the total dosage used is varied with the density of the tussock moth population. However, they commonly use 76,800,000 particles per hectare, or 18 larval equivalents. There is a lapse of 5 to 7 days before death occurs. After about 10 days, the mortality is about 50%, and a complete kill is usual after 14 days. However, hard rains reduce the effectiveness. Leaders are ready to popularize the use of the virus. Workers have been taught how to produce it and are now doing so.

Scientists have studied the characteristics of the virus. The size of a particle is 1.25 × 3.75 microns. Each particle has 6 to 10 rods, and even up to 28. Each rod becomes rootlike when the outer coil is dissolved. The dimension of the virus is 60.5 to 375 millimicrons. It may belong to the genus *Borrelina*, but it has the shape of another genus (sausage shaped). It is now placed in Group I of the baculoviruses.

At the Long March (Changchen) People's Commune in Shanghai, which has 29,000 people and 1,055 hectares of cultivated land, cutworms, especially *Agrotis ypsilon*, are a problem. Five control methods are used: (1) vinegar bait with a toxic agent for the moths; (2) flooding at a proper time to drown the larvae; (3) fertilization to improve soil adversity (it was not explained how); (4) use of blacklight traps, one per 5 acres; and (5) use of *Bacillus thuringiensis*. The commune has a unit for producing its own *B.t.*; it uses a wheat bran-cornmeal mixture and inoculum supplied by the province. Test tubes are inoculated and the culture is fermented for 24 hours, i.e., to the "white solid state." The culture is then diluted by the wheat bran-cornmeal mixture. After an undisclosed time interval, they assess for spore density and again dilute 700 to 800 times to give a spore count of 10 billion per gram of product, which is applied in the field. It was off-season for this work, and we did not see it, but we did get a photograph of a small storage shed where stock material was stored in bottles about 1 liter in size.

The Tahsia (Big Sand) Commune in Kwangtung Province has its own *B.t.* production facility. This microbial material is used against a number of rice pests in the province. They use a fermentation tank method to produce a product which, for the most part, seems to be used as the inoculum to greatly augment the production by use of a commune-developed solid-production medium using rough mixing and storage facilities (a shed or small room). The latter method requires less labor and no special industrial facility, and it is much cheaper. A quality-control assessment is made of the product to be used for inoculum.

A peanut bran and water nutrient is used in the fermentation method; fermentation is for 30 hours, after which the product is concentrated by passing the liquid through filters under high pressure.

The filtrate (the product) is then spread out and dried on racks in the sun and powdered in a mill, and placed in plastic bags. For the solid medium production, they had on the floor, in a rough shedlike room, a pile of peanut bran, a pile of soybean meal, and a container of inoculum. The first two were shoveled into a large shallow basket 2 feet in diameter, and a measured amount of inoculum was added. The lot was then mixed by hand. Some of the workers used a breathing mask; others did not. This medium, kept just damp, was also cultured for 30 hours and was said to be as effective as a factory product.

The material is bioassayed for effectiveness at the brigade level
before use.

Use of Virus against the Cotton Leafworm (armyworm) *Prodenia litura* Fabricius

Researchers in Chungshan University in Canton have been studying a
nuclear-polyhedral virus (NPV) of the cotton leafworm, *P. litura*. This
is a pest of various vegetables and legumes in the suburbs of Kwangchow,
Kwangtung Province. In the past, it has been controlled by insecticides.
An NPV that had been reported in Egypt was found in Kwangtung Province
in 1960 (Tai, 1973)[18]. Research on it offers hope that use of the virus
will reduce the use of insecticides "...as the people eagerly wish,"
it was said. Promising field results have been reported by some brigades (over 80% kill).

A chart was shown giving the characteristics of the virus and details of its production, dosage, and application.

Miscellaneous Biological Control Efforts

Although many efforts to expand the use of biological control are probably under way in China, our group saw only a few examples. These included various uses of entomopathogenic microbes, use of these agents
and various parasites and predators in a number of integrated control
efforts, and a few efforts against unusual or speciality-crop pests,
such as the litchi stinkbug, a noctuid (*Eublemma amabilis*), which is a
serious pest of the lac culture, and the mulberry tussock moth.

Biological Control of a Noctuid Pest of the Lac Industry The purple
lac insect (*Tachardia lacca* Kerr), is cultured for its shellac on forest trees of *Dalbergia balansae* and other Papilionaceae in Kwangtung
Province in southern China. Suitable trees are selected in the forest,
and infestations of appropriate densities are initiated on the branches.
A noctuid, *E. amabilis*, destroys the lac and probably feeds directly on
the lac insect itself, it was said, and is thus considered a "primary
natural enemy" of the lac insect. Each larva destroys 2-6 cm of lac.
If the noctuid reaches a density of 100 larvae per meter of "stick,"
most of the lac is destroyed. If such a density occurs at an early
nymphal stage of the lac insect, the crop is lost. If the lac insect
is in a mature stage, both yield and quality are affected. Chemical
control of the noctuid is very expensive and not feasible in the moun-
tainous terrain.

P. S. Negi and associates in India reported in 1946 on the possibilities of biological control of this pest by use of the parasite,
Microbracon greeni (Negi *et al.*, 1946)[13]. Chinese scientists of the biological control group in Kwangtung searched on Hainan Island and brought
the parasite from there to Kwangtung Province in 1972. The parasite
was readily established.

M. *greeni* is produced in cultures on *Eublemma* itself as the main insectary host. But the pink bollworm is also a good host for rearing for augmentative release purposes. The rate of release of M. *greeni* is determined by the density of E. *amabilis*. If 10-30 per square meter are found, 180 to 300 female M. *greeni* are released per acre of lac forest. The sex ratio of M. *greeni* is three females to one male. The rate of release is increased if the density of E. *amabilis* is higher. The parasite was found to spread readily.

M. *greeni* has proved capable of effectively controlling E. *amabilis* in a short time. The parasitism in release areas has been generally about 60%, with a maximum of 90%. The parasite lowered the pest insect density to five insects per meter of lac stick (sometimes one insect), and thus increased production of lac by 20%-40%. Also, the lac produced was of much better quality than that produced in control areas.

Since the lac harvesting process nearly annihilates the lac habitat and also E. *amabilis*, the M. *greeni* population in the forest is also nearly destroyed, and M. *greeni* is not very effective in winter. It is thus necessary to maintain high densities of M. *greeni* in the habitat. One way of doing this is to add the pest species itself (*Eublemma*) into the lac area (about 1,536 per hectare) at times of scarcity of *Eublemma* for B. *greeni* to attack. This is also done following low winter populations, so that continuing pressure by the parasite can be maintained. In the critical situations, pest-augmented areas had parasitism of 54% while nonaugmented areas had parasitism of only 7.4%.

This example of obtaining better pest control by artificially introducing (augmenting) the pest population at critical times in order to ensure adequate biological control has been shown to be technically feasible for control of cyclamen mites on strawberries in California[7] and for control of the cabbage butterfly on cabbage in Missouri[15]. But in neither case have those methods been adopted by farmers. The Chinese socialist system expects soon to popularize this method for control of the pest of lac and to use it in all the lac-producing area. If this is done, it will be notable first in general application. If E. *amabilis* is a true predator of the lac insect (i.e., if it seeks it out and does not destroy it only incidentally), this would be a clear example of the use of a secondary natural enemy to control a primary natural enemy.

Biological Control of the Litchi Stinkbug In Kwangtung Province, the litchi stinkbug is a serious pest of litchi fruits. Scientists at Chungshan University have collaborated with a litchi-producing commune to develop a program of biological control of this pest. They mass-produce the eupelmid egg parasite, *Anastatus* sp., on eggs of the giant silkworm, *Samia*, in a manner similar to that described for *Trichogramma* production. The advantages described for production of *Trichogramma* apply in this case as well.

In order to maintain good production, properly synchronized at the time of need against the stinkbug in the field, they obtain good parasitism of the eggs of the giant silkworm by *Anastatus* in the fall of the year and place them indoor conditions to overwinter. At the proper time in the spring, they are brought back into the outdoor insectary where the adult parasites emerge in several days, depending upon the temperature. They are then exposed to fresh batches of giant silkworm eggs, which they parasitize. The prepared parasitized egg sheets are then placed by hand as needed in the litchi trees. An illustrated bulletin was produced as part of the popularization of the program.

The biology and methods of propagation were studied in 1962 by research personnel. In 1966 and 1967, research and extension people cooperated in large-scale demonstrations of field operations in two counties in Kwangtung Province. Good results were obtained. In late 1969 and 1970, biological control technicians were trained, and in 1970 they cultured and released the wasps in 12 communes. Parasitism reached 85.5%-98.7%, compared with 10%-14% in nonrelease areas. In 1973, there were over 50 wasp-rearing stations in the 11 counties and municipalities in Kwangtung where litchis are grown (Huang et al., 1974)[6].

Biological Control Agents in Integrated Control of Rice Pests

Because it became obvious that in Kwangtung Province biological control agents could not be depended on to adequately control all the insect pests of rice, an experimental program of integrated control was developed. This includes use of several biological control agents used in conjunction with other measures: (1) cultural methods, including use of trap crops, drowning of immature stages, and so on; (2) accurate forecasting of pest densities; (3) use of blacklight traps (one per 2.4 hectares); and (4) use of chemicals at least in part at reduced dosages either on early-planted trap-crop areas only or mainly in high-density spots in the fields (these were treated with methyl parathion where needed). The biological control agents included: *Trichogramma*; *Bacillus thuringiensis* (in combination with low dosages of dimethoate); use of ducks for insect control (and probably serving somewhat also for weed control and mosquito control, and possibly for snail control); protection of frogs as useful insect predators; and use of a rice variety resistant to a sclerotizing fungus-disease pathogen, *Rhizoctonia*. The combined measures in one experimental program reduced the yellow rice paddy borer from about 4% to 1% (and less), reduced the insecticides used by about 66%, and reduced the cost of insecticide use from 20 yuan per hectare in 1974 to 6 yuan per .39 hectare in 1975. It was reported that this system of integrated control will soon be tried by other communes.

Ducks used as biological control agents are a most interesting development in Kwangtung Province (Figures 7 and 8). In the above experimental work, 3,500 ducks were used at an average of 12-15 ducks per .39 hectare (2 or 3 per mu), but they are "herded" through small paddies in flocks of 1,000 to 1,200. This is an old Chinese practice and was pushed by the commune workers as an alternative to the use of hazardous insecticides. The ducks also furnish meat and income. The potential great value of ducks for control of rice pests (both insects and weeds) and of vectors or alternate hosts of human disease (mosquitoes, and snails, with reference to malaria and schistosomiasis) was

FIGURE 7 A public display
showing how ducks can be used
for insect control in rice
fields.

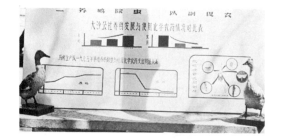

given little concern by our hosts. This may be because both malaria
and schistosomiasis are no longer said to be problems in the area.
    The Tahsia (Big Sand) Commune in Kwangtung Province (3,900 hectares)
is already practicing a substantial integrated-control program for
their rice pests, with technical and scientific collaboration by the
staff of Chungshan University. The program in this three-crop rice
area embraces most of the above biological control items and other as
well. Formerly, the commune used only insecticides, and although they
got good pest control, it was expensive and required heavy treatments.
The chemicals also produced outbreaks in some species. Integrated control is now used on all their rice acreage. Cultural control is considered the base. They have their own program of production of
Trichogramma and *Bacillus thuringiensis*, and in 1975 they raised
220,000 ducks, which were used, in large part at least, in pest control
in their early rice crop. Although they are fed some grain products
(e.g., bran) twice a day, the ducklings are kept hungry enough to forage avidly during the day. As the rice matures, the ducks grow larger
and can reach higher up on the plants to feed on the insects. Presumably, smaller numbers of the larger ducks are then used at a given time.
Previously, the Chinese used only insecticides for control of important
planthoppers, *Nilaparvata lugens* Ställ, *Niloparvata oryzae* Matsumura,
Sogata furcifera Horvath, and *Laodelphax striatella* Fallen. In 1975,
they used only ducks for these pests. On June 16, there were 264-336

FIGURE 8 Baby ducks in a rice
field. The ducks consume
insects in large numbers.

planthoppers per 100 rice plants. But by June 21, the ducks had reduced the populations to 80-120 per 100 plants, a reduction of 65%-
70%. Sometimes the populations surpass safe levels, and then
insecticides are used in the indicated areas only. One duck was
observed to eat 274 insects in 2 hours; 80 were natural enemies and
194 were harmful species (not necessarily a beneficial ratio, it would
seem). In a test, one duck ate 69 rice grasshoppers in 2 hours, and
another ate 417 insects of various types in 2 hours. The use of ducks
has made it possible to cut the normal weed-control effort in half.

In this commune, *Bacillus thuringiensis* is used for its selectivity;
it is nonhazardous to the workers and to *Trichogramma* and other natural enemies (including frogs and ducks). Chemicals are used selectively
(by the dosage, area treated, and timing), as necessary. *B.t.* gives
about 75% kill of the rice leafroller, and it is used for armyworm control with good results. For the third and sixth generations of leafrollers, they release 60,000 to 600,000 *T. australicum* or *T. dendrolimi*
per .39 hectare and get about 75% parasitization. This year (1975) they
are trying *T. japonicum* (suggested as the best one for this area) and
are releasing only 60,000 to 300,000 per .39 hectare.

Biological Control Agents in Integrated Control of Citrus Pests

In Kwangtung Province, we were not informed of an overall integrated-
control program on an agricultural production basis for citrus pests.
However, since 1973, studies have been initiated. For a year, an
experiment has been conducted in a 7-year-old citrus grove. It concentrates on biological control of citrus mites: *Schizotetranychus
sexmaculatus*, *Phyllocoptruta oleivora*, and *Panonychus citri* (McGregor).
Entomologists are studying the effects of *Stethorus* spp. and phytoseiid
mites as predators.

In this experiment, they protect the natural enemies of all citrus
pests by using types of insecticides having some selectivity, including the botanicals. The spider mites (above) and the rust mite are
said to be controlled at densities of less than one mite per leaf by
the phytoseiid *Typhlodromus victoriensis*. Since April (1975), no
insecticides or acaricides have been used. The scale insects are said
to be adequately controlled by hymenopterous parasites. Included are
the pest scales *Chrysomphalus ficus* (Ashmand), *Parlatoria pergandii*
Comstock, *Parlatoria zizyphus* (Lucas), *Unaspis yanonensis* Kuwana,
Lepidosaphes beckii (Newman), and *Aonidiella aurantii* (Maskell) and
a whitefly, *Aleurolobus marlatti* Quaintance. They did not specify
which parasites are effective on each of these pests, but said that
Aphytis chrysomphali, *Aphytis mytilaspidis*, and *Anagyrus aurantifrons*
are involved in part. There is a leaf miner, *Phyllocnistis citrella*,
for which they do not have adequate biological control; a selective
material is used against it. They do not know how to control Lepidoptera during the flush growth period.

In Hunan Province, extended effort over a number of years has been
devoted to integrated control of citrus pests, including much attention
to biological control (Table 3).

TABLE 3 Citrus Scales in Hunan Province and Their Parasites

Prontaspis yanonensis	Parlatoria pergandii	Parlatoria zizyphus	Chrysomphalus ficus
Aphytis chrysomphali	Aphytis mytilaspidis	Aspidiotiphagus citrinus agilior	Aphytis chrysomphali
A. proclia	A. proclia	Aphytis proclia	A. maculicornis
A. aonidae	A. aonidae	A. aonidae	Comperiella bifasciata
	Aspidiotiphagus citrinus agilior		

Although cultural management is said to be the foundation of citrus pest control, biological control probably keeps many potentially serious species at low numbers, e.g., those scale insects listed in the following table except for *P. pergandii*. The "red lady beetle" (presumably the vedalia, *Rodolia cardinalis*) was introduced for control of the cottonycushion scale, *Icerya purchasi*.

Biological Control Agents in Integrated Control of Pests of Pome and Stone Fruits

At the Northwest Agricultural College in Wukung, Shensi Province, Chinese scientists have been conducting a significant study to evaluate pests of deciduous fruits and "accepted" practices of pest control in orchards.

There are some 120 species of pests or potential pests of these crops, and about 20 cause consistent damage. Among these are *Choristoneura lonicellana* (Walsingham), *Leucoptera scitella*, the spider mites *Tetranychus viennensis* Zacher and *Bryobia praetiosa* Koch, San Jose scale, *Quadraspidiotus perniciosus* Comstock, oriental fruit moth, *Grapholitha molesta* Busck, and a number of diseases.

In the early years, few pesticides were used and the pests were not under good control. In 1950, the main pests of apple were leafrollers, e.g., *Adoxophyes orana*. The pests overwinter under the bark of the tree. Use of DDT at budding time was successful for a few years, but the sprays killed many natural enemies. By the early 1960's, most of the natural enemies of spider mites (e.g., *Stethorus punctillum*, *Chrysopa septempunctata*, and *Scolothrips* sp.) had been largely eliminated. Spider mites then became the most important pests of apple in the province. There are five species of leafrollers (*Adoxophyes*, *Choristoneura*, *Leucoptera*, *Illiberis*, and *Spilonota*) whose combined parasitization is usually 25%-27%.

Chinese scientists have recently been spraying about 10 times a year, mostly for spider mites; two or three applications are for diseases. Spider mites have become highly resistant to the organophosphate materials, e.g., tedion and parathion. As a spray for San Jose scale, they use lime-sulphur and soda water with a detergent. This is used because it is not so destructive to natural enemies, such as *Aphytis proclia*.

The Chinese found that where broad-spectrum insecticides were not used, *C. septempunctata* destroyed 38%-50% of the brood of *L. scitella*. A variety of parasites were also effective, killing about 25% of the overwintering pupae. In the fourth generation of the host, they destroy about 74%. But when the highly toxic materials were used, *Chrysopa* disappeared and parasitism dropped to 1.6%-6.7%.

On peaches, parathion was used in July and August for control of the oriental fruit moth. This destroyed the effective natural enemies of San Jose scale, especially *A. proclia*, and outbreaks ensued. Many other unwise practices were encountered, practices in which destruction of natural enemies occasioned outbreaks by previously nonserious pest species. Now insecticides are no longer used as the single method. Promising results have been achieved in some respects from a combination of methods. These methods include pruning; removing scaly bark; clearing away deadwood, fallen leaves, apple mummies, and other crop residues; using chemicals selectively; and using blacklight traps and sugar-vinegar baits.

Biological Control Agents in Integrated Control of Cotton Pests

No attempt is made here to detail all the biological control elements in the integrated control of cotton pests in China.

Use of *Trichogramma* has been discussed.

In Peking, we were told that *C. septempunctata* is widely transferred from natural vegetation or other crops into cotton for control of cotton aphids. This is being practiced on about 19,500 hectares of cotton in various provinces. In the Shanghai area, and also in Shansi Province, one biological control element that is used in an apparently successful integrated-control program for the pink bollworm in cotton is the parasite *Dibrachys cavus*, which is used primarily in the seed-storage houses. Blacklights are also used in the storage houses. These methods are used to reduce the infestation sources in the seed supply. The pest pupates in the seeds. In the Shanghai area, chickens are used to feed on the larvae leaving the seeds when they are spread out in the sun. We did not learn whether they have a rearing program to produce the parasite, but we saw a wall chart suggesting that they have one.

In the Tangwan People's Commune in Shanghai, the delegation was told that spider mites are not a problem on cotton or vegetables. Our hosts considered (without direct evidence) that their irrigation and fertilizer programs were a reason for this, but they said that spider mite predators might be a factor. In the Long March Vegetable (Changchen) Commune, however, they stated that spider mites are the main problem, for which they use dimethoate. At first, cutworms and aphids were treated with chemicals. Then spider mites became worse, possibly because of the increase in variety of crops being planted or because of destruction of spider mite enemies.

At the People's Evergreen (Szechiching) Commune in Peking, vegetables are grown extensively in greenhouses. In this greenhouse there is a significant spider mite problem, for which dimethoate and dichlorvos are used. It is surprising that we saw no indication of experimentation with releases of *P. persimilis* for control of spider mites in greenhouses. This practice is widely used in the U.S.S.R., Finland, and some other European countries (Hagen *et al.*, in press)[4].

Dimethoate is used for cotton aphid control in the Shanghai area, although it is not reported to be effective in U.S. cotton. Its use is known to kill natural enemies of the cotton aphid and of bollworms, and to be associated with increases in bollworms (e.g., in California). Some important natural enemies of the aphid in the Shanghai area that are affected by chemical treatments are coccinellids, *Chrysopa*, and a braconid (not named). (*Adalia* sp. feeding on aphids on crape myrtle were collected near an aphid-infested cotton field.)

Conclusion

The group's visit to China was very worthwhile. While we perhaps did not gain much pioneering information at the basic science level, the experience caused many of us to take a new look at some of the attitudes and accepted practices concerning various aspects of insect pest control in the U.S. For example, seeing the vast extent and successful use of cultural control measures as practiced in China serves far better than anything available in the U.S. to impress us with the need to pursue this area with greater vigor. An additional example is offered in the very wide use in China of *Trichogramma* for biological control of various crop pests. The prevailing attitude in the U.S. has been that use of *Trichogramma* is not a promising tactic in the U.S. The impact on this attitude of what we observed in China suggests that the subject needs more careful research and examination.

REFERENCES

1. Agricultural Experimental Station, Sanke-yushu People's Commune, Tung Hua County, Kirin Province. 1975. The control of European corn borer by using trichogrammatid egg parasites. Acta Entomol. Sinica *18*:10-16.
2. Collaborative Research Group on Biological Control of Rice Pests, Kwangtung Province. 1974. The control of rice leafroller, *Cnaphalocrocis medinalis* Guenee, by trichogrammatid egg parasites. Acta Entomol. Sinica *17*:269-280. [with English summary]
3. Division of Insect Resources, Kwangtung Institute of Entomology. 1975. Experiments to control *Eublemma amabilis* Moore by liberating *Bracon greeni* Ashmead in the forests of lac production. Acta Entomol. Sinica *18*:141-150. [with English summary]

4. Hagen, K. S., S. Bombosch and J. A. McMurtry. In press. Biology and prey relationships of predators, Chapter 5. *In* "Theory and Practice of Biological Control," C. B. Huffaker and P. S. Messenger, eds. Academic Press, N.Y.

5. Hsiu Cheng-fung, Chang Yung, Kwei Cheng-ming, Han Yu-mei and Wang Hwei-hsien. 1973. Field application with *Beauveria bassiana* (Bals.) Vuill. for European corn borer control. Acta Entomol. Sinica *16*:203-206.

6. Huang Ming-dau, Mai Siu-hui, Wu Wei-nan and Poo Chih-lung. 1974. The bionomics of *Anastatus* sp. and its utilization for the control of lichee stinkbug *Tessaratoma papillosa* Drury. Acta Entomol. Sinica *17*:362-375. [with English summary]

7. Huffaker, C. B. and C. E. Kennett. 1956. Experimental studies on predation: (1) predation and cyclamen mite populations on strawberries in California. Hilgardia *26(4)*:191-222.

8. Hwang Guan-Huei and Ding Tsuey. 1975. Studies on the nuclear polyhedrosis-virus disease of the cotton leafworm, *Prodenia litura* F. Acta Entomol. Sinica *18*:17-23. [with English summary]

9. Klassen, W., ed. 1973. Biological methods of protecting fruit and vegetable crops from pests, diseases and weeds as bases for integrated systems. [Summaries of Reports, All-Union Scientific Research Institute of Biological Methods of Plant Protection, Kishinev, USSR, Oct. 1971.] ARS, USDA. 173 pp. Mimeo.

10. Kwangsi Kweishien Biological Control Station and Laboratory of Plant Protection, Kwangsi Academy of Agriculture. 1974. Experiments on the integrated control of rice insect pests, chiefly by parasitic wasps, bacteria and fungi. Acta Entomol. Sinica *17*: 129-134.

11. Kwangtung Institute of Entomology. Biology Department, Chungshan University. 1973. Use of *Anastatus* to control lichee stinkbug. Kwangtung People's Press, Kwangtung. 40 pp.

12. Laboratory of Biological Control, Department of Plant Protection, Kwangsi College of Agriculture. 1974. Investigations on the techniques of utilizing of egg parasites, *Trichogramma* spp., to control insect pests. Acta Entomol. Sinica *17*:258-268.

13. Negi, P. S., S. N. Gapta, M. P. Misra, R. V. Venkataraman and R. K. De. 1946. Biological control of *Eublemma amabilis* Moore by one of its indigenous parasites, *Microbracon greeni* Ashmead. Indian J. Entomol. 7:37-40.

14. Pang Xion-fei and Chen Tai-lu. 1974. *Trichogramma* of China (Hymenoptera: Trichogrammatidae). Acta Entomol. Sinica *17*:441-454. [with English summary]

15. Parker, F. D. 1971. Management of pest populations by manipulating densities of both hosts and parasites through periodic releases, pp. 365-376. *In* "Biological Control," C. B. Huffaker, ed. Plenum Press, N.Y.

16. Seed and Plant Protection Station, Tung Hua County, Kirin Province. 1975. The experience and realization of large scale control of European corn borer by using trichogrammatid egg parasites. Acta Entomol. Sinica *18*:7-9.

17. Sha Cha-yun, Ren Gai-xin and Xie Qiang-jiang. 1975. Studies on phage-resistant strains of *Bacillus thuringiensis* var. *gallerie* "010" and the phages of *Bacillus thuringiensis*. Acta Entomol. Sinica *18*:273-280. [with English summary]
18. Tai Guan-Chuen. 1973. A preliminary study of the polyhedrosis on the cotton leafworm *Prodenia litura* Fabricius in Canton Area. Acta Entomol. Sinica *16*:89-90.

C. HOST RESISTANCE

General

Breeding for resistance to insect pests is in its infancy in China. We visited only two locations having any significant activity in this field. No work was evident at places visited in Peking, although no opportunity occurred to visit the Institute of Genetics. Entomologists at the Peking Institute of Zoology indicated that no cooperative work in the field had materialized in the Peking area but expressed interest in doing future research. Separation of disciplines--entomologists at the Peking Institute of Zoology and plant breeders at the Institute of Genetics--handicaps future research.

Our first evidence of a program in host-plant resistance to insects came at the Kirin Academy of Agricultural Sciences at Kungchuling, Kirin Province. At this station, there is a Plant Breeding Institute, a Soil Fertilizer Institute, and a Plant Protection Research Institute, where the entomologists were located. This arrangement permits close contact between the disciplines. Corn, soybeans, and sorghum were the predominant crops at this location. We met with three plant breeders, representing small grains, corn, and soybeans, to discuss their breeding programs and principal pest situations on the major crops. The scientists are not conducting any plant-resistance work on corn and small grains, although they are making some observations in the field and expressed hope of doing more in the future. Although the European corn borer, *Ostrinia nubilalis*, is a major pest in this area, no selection and evaluation work was being done with respect to this insect. They indicated interest in synthetics rather than hybrids and were using a broad base of germ plasm and making selections. We told them of the extensive program in the United States designed to develop resistances to the European corn borer and left several lines with them that carried resistance in the United States to the European corn borer and certain diseases.

Major soybean insect pests in China are the soybean pod borer and aphids. The soybean pod borer normally infests more than 30% of the pods if not controlled. Primary control consists of cultural methods, resistant varieties, and chemicals when needed. In Kirin Province, there is only one generation of the soybean pod borer. In mid-August the adult emerges and lays eggs on leaves and bean pods. It was observed over several years that many of the young larvae failed to penetrate the pods of several varieties. Entomologists were not sure what mechanism of resistance was present. They think that the pod toughness and possibly increased pubescence may be factors involved. Antibiotic factors in the carpel wall of the pods may also be present. Although several lines were identified through field screening, two varieties, Keling #3 and Tieh-chia-shi-li-huang, have been identified as having moderate resistance (about 50%). Keling #3 is considered the better of these two varieties and has been popularized in the area. We obtained seed of this variety. We asked how many hectares of Keling #3 are currently cultivated in the province. We received no definite answer, but the plantings were said to be very extensive, particularly in the Kungchuling area. Soybean varieties were being evaluated under field conditions for aphid resistance, but no resistant variety had been found.

There was no planned program in insect resistance in small grains and sorghum. The Chinese indicated that they are hoping to emphasize this activity in the future. One of their major pests is the greenbug, *Toxoptera graminum*, and we indicated to them that several sources of resistance had been identified and utilized in the United States in both small grains and sorghum. We presented a number of U.S. wheat lines containing resistance to the Hessian fly, various plant diseases, and drought, along with several sorghum lines containing resistance to the sorghum midge and the greenbug.

There is considerable emphasis on breeding for resistance to all major plant diseases occurring on corn, soybeans, wheat, barley, and millet in the area. This work was discussed in the report by the plant science group who visited the station last year, so is not covered in this report.

While in Kirin Province, we also visited the Kirin Institute of Forestry Research, the primary forestry research unit in China. Some interesting work in biological control of the pine moth with *Trichogramma* is being conducted. But there is no activity in host-plant resistance, nor is there any indication that such activity will be started soon.

The greatest activity in breeding for resistance to insect pests is in the Wukung area, Shensi Province, at the Northeast College of Agriculture (about 144 km west of Sian). Activity is with cotton and wheat, which are important in this area.

Cotton

A study was begun in 1973 to select for resistance to the cotton aphid, *Aphis gossypii* Glover. The cotton aphid is one of the most harmful cotton pests in this area, requiring a number of insecticide applications.

The Chinese scientists identified nine upland varieties and five Egyptian varieties as having resistance. The delegation was told that 385 North is the most resistant and is rapidly being popularized in the area. Aphids have a lower reproduction rate and develop more slowly on this line. The aphids also are restless, become scattered, and feed less. The primary mechanism of resistance is apparently antibiosis. Shensi and Far North 1 were two other varieties mentioned that contained some resistance. We were not able to obtain seed.

The Chinese estimate that 700,000 mu (47,250 hectares) of aphid-resistant cotton is being grown in Shensi Province alone, and the variety has been spread into adjoining provinces. An accurate estimate of total acreage could not be made. Popularization is continuing at a rapid rate because of the severity of the aphid problem and mounting insecticide resistance in the cotton-producing area.

Wheat

Wheat midges are serious pests in this area, especially following heavy rains. Two species are involved: *Sitodiplosis mosellana* (Gehin) and *Contarinia tritici* Kirby. They have been basically controlled since 1960 through a combination of a resistant variety, soil treatment, and protection of natural parasites.

The resistant variety of wheat was developed through selection from survivors of a native variety in a heavily infested area. A resistance mechanism is the glume, which is tightly closed. The angle of the glume (outer/inner) is also important; so is the hard, plasticlike texture. The hard texture allowed for a higher fertilization rate without damage. The ovary and seed grow very fast--another resistance mechanism. Resistant characteristics were carried by a dominant gene. It was indicated that large plantings of the resistant variety had nearly eradicated wheat midge populations.

Insects and mites transmitted five major viruses in the area: barley yellow dwarf, north cereal virus, wheat blue virus, winter wheat virus, and wheat streak virus. Barley yellow dwarf, the most serious, was transmitted by the greenbug and other grain aphids. The north cereal virus was transmitted by planthoppers, and wheat streak virus by spider mites. Aphids transmitting barley yellow dwarf virus also caused serious physical damage and were widely dispersed. Aphid species differed in various areas, but the most important species was *Toxoptera graminum* (Rhondani).

The virus complex was controlled by the following practices:

- Forecasting aphid presence in the three-leaf stage of grain. This is based on population numbers, incidence of virus (greenhouse assay), dispersal, climatic factors, number of natural enemies, and variety planted.

- Planting of resistant varieties selected from native wheat strains.

- Some insecticide, including seed dressing with 0.3% of phorate or disulfoton, which controls soil insects and aphids for 30 days.

In the Shanghai area and in the great rice-producing area of Hunan and Kwangtung Provinces, no organized research or utilization program is evident in host-plant resistance to insects. This seems particularly unfortunate since good sources of resistance to leafhoppers and rice stem borers have been identified at the International Rice Research Institute in the Philippines and at other rice research stations. Rice insects are major problems in southern China, and resistant varieties would offer an ideal way to solve some of them. Breeding emphasis is still on developing short, fast-maturing *indica* varieties that are suitable for double- and triple-cropping. The Chinese have used traditional varieties to select desirable characteristics, such as early maturity and short stem. Perhaps this has the advantage of carrying along greater tolerance to native insect pests and diseases, and it may be preferable to introducing narrow gene pool foreign germ plasm that may contain no resistance to China's local pests.

Real opportunity exists for exchange of rice germ plasm containing resistance to various pests. There is considerable concern that as the Chinese develop new varieties and replace old ones, there may not be a mechanism for preserving seed of the old varieties. It is possible that much of the extremely valuable old rice germ plasm cultured in the great rice basin of China for thousands of years could be lost to the world forever.

Probably the most fruitful area for both the United States and China in future agricultural-resource exchanges is the area of crop germ plasm. China could benefit immeasurably from an effective exchange program, taking advantage of U.S. breeding progress for resistance to major insects and diseases in corn, sorghum, wheat, soybeans, legumes, and truck crops. The United States could benefit by obtaining soybean, sorghum, wheat, millet, barley, and rice germ plasm. The Chinese grow cotton in regions farther north than do U.S. producers. There is a current emphasis on developing short-season cottons in the United States to escape insect infestations and reduce energy inputs. These short-season varieties, adapted for the more northern cotton-growing areas of China, may be valuable as a germ plasm source for U.S. breeding programs.

China, in order to make another giant stride forward in agriculture, undoubtedly needs to emphasize crop variety improvement, especially in relations to insect and disease resistance, yield, and quality. If China does emphasize such a program in the years ahead, she may wish to exchange information and germ plasm with the United States and other nations. She could thus continue to show the world her strength in agriculture by feeding a nation of about a billion people while producing food and fiber surpluses for world market trade.

D. SEX PHEROMONES, HORMONES, AND TRAPPING

Sex Pheromones

Sex pheromones have become the glamour area of entomology during the past decade in many countries of the world, notably in the United States, Australia, Germany, Japan, Holland, Switzerland, England, and Canada. These behavior-modifying chemicals, with their potential for insect control, have been intriguing to both scientists and the public. Rapid progress has been made on the identification of chemical structures, which has provided the basis for other academic studies on behavior and sensory physiology. More recently, many countries have supported practical research directed toward the use of pheromones for monitoring the presence and abundance of insects and for direct control. Although only a small number of investigators throughout the world have been pursuing the practical research for a short time (about 5 years) with limited resources, the glamour has faded somewhat as the anticipated panacea has not developed rapidly. However, pheromones have found important commercial use in pest management programs, and research on their use in insect control is still showing encouraging progress. Not discouraged by the amount of time required to put pheromone use into practice, Chinese scientists have entered this area with great interest and government support.

Some research had been initiated before the summer of 1973, but much of the interest apparently has been developed since then by the scientists who visited many U.S. pheromone research laboratories as members of a Chinese Insect Hormone Delegation to the United States June 25-August 6, 1973. These scientists now play key roles in programs designed to bring pheromones into practical use. The widespread interest in sex pheromones throughout China reveals that these programs have been taken seriously by both scientists and administrators.

The expertise for identifying new sex pheromones has been developed by a number of investigators in various countries, but it is still difficult to initiate these studies "from scratch." Pheromone research in many countries, therefore, consists in applying the already reported basic chemicals and biological knowledge to species that have been investigated elsewhere. To some extent, this is true in China, but there is basic research on the identification of the specific pheromones of some of the major Asian pests. The most advanced research group in this respect is found at the Peking Institute of Zoology. A group at this institute was one of the first to conduct research on pheromones in China, and by 1973 they had already published their preliminary findings of research conducted in 1966, 1967, and 1972 on the pheromone of the pine caterpillar moth, *Dendrolimus punctatus* (Walker)[1].

In their recent studies on pheromone identification, they have incorporated the electroantennogram technique[2,3] into their research program (Roelofs and Comeau, 1971a,b). In a unique and simple setup, the ground electrode (tungsten wire) is positioned in a piece of wet cotton immediately beneath the isolated antennal base. The antennal distal tip is inserted (or sucked with a small rubber bulb at the other end of the capillary) into a small capillary containing saline solution and the input electrode (tungsten wire). The two electrodes are rigidly fastened together so that the whole antennal setup can be moved at random without disturbing the antennal hookup. A homemade 100X amplifier feeds the signals into an oscilloscope. A short glass-tube cartridge containing test material on filter paper is fitted into a side-arm glass joint positioned perpendicularly to the airstream, and air from a glass syringe is puffed through the cartridge to assay antennal responses. A demonstration with extract from female pine caterpillar moths showed that the male antennae give very good EAG responses to the extract.

Pine caterpillar moth Dendrolimus punctatus Pine caterpillar moths are major pests of pine trees in the eastern, southern, and southwestern parts of China, with three or four generations occurring a year. There are at least 11 species of *Dendrolimus* in China: *D. biundulata, D.*

densate, D. himalayanus, D. kikuchii, D. latipennis, D. omeiensis, D. punctatus, D. sibiricus, D. superans, D. tabulaeformis, and *D. xichangensis. D. punctatus* became the target for pheromone research because of its major pest status. It could be a complicated example because there are 2 subspecies: *D. p. techangensis* and *D. p. wenshanensis.* If the subspecies use the same pheromone, however, the problem would be obviated. The identification of a pheromone of one of the pine caterpillar moth species would open the door for interesting cross-specificity and reproductive-isolation research among the various closely related species, and would provide a tool for monitoring and control.

Field trapping results from the preliminary studies[1] in 1966 and 1967 were obtained by cooperating with workers in the Stone Gate Mountain Forest Experiment Station, Anhwei Province (two provinces south of Peking), and in 1972 by cooperating with workers in the Anhwei Production and Construction Division, the Shachun Forest Experiment Station, Kasun County, Anhwei Province, and with the Agricultural Science Research Institute in Kiangsu Province (on the coast immediately east of Anhwei Province).

The first task was to prove that the insect actually uses a pheromone communication system for mating. Cardboard cylinder traps (20 x 8 cm) were each baited with a virgin female. One end of the trap had a funnel entrance with a 1.5-cm hole in the center; the other end was covered with transparent paper. In five trials, 16-48 traps were placed in the pine forest, and in each trial they captured 1.0-1.4 males per trap per night, compared with 0 in 20 unbaited traps. Between one-third and two-thirds of the baited traps caught males each night. These results indicated that a pheromone system is used by the pine caterpillar moth. It was reported that the attractiveness of the pheromone varied (1) with environmental factors such as wind velocity, temperature, amount of sunlight, and the height and density of the trees in the area; (2) with trap factors, including the type of construction, the height of deployment, and the direction in which the traps are positioned; and (3) with insect factors, such as the density of females per tree and individual variations in pheromone production. The time of mating activity was then determined by observing the male catches at 0000, 0100, 0330, and 0530 hours for three nights. Almost all of the males were caught after midnight between 0100 and 0530 hours.

The results of the next experiment to determine if the female produces pheromone only at the time of "calling" have very interesting implications. A crude extract of 700 female abdominal tips was made at 2100-2200, at 0000-0100, at 0200-0300, and at 0400-0500. Each extract was then dispensed onto 30 pieces of filter paper, which were used individually to bait cylinder traps. In 4 nights of field trapping, the 30 traps containing extract obtained at 2100-2200 captured only 4 males, and the traps containing extract made at 0000-0100 caught only 8 males. Meanwhile, the extracts obtained from presumably calling females at 0200-300 caught 194 males, and the extract made at 0400-0500 caught 442 males, compared with 0 in 13 unbaited traps. These data indicate that the females are producing the extractable pheromone only at the time of calling and that the extracts should be obtained at that portion of scotophase, preferably between 0200 and 0430.

An investigation of extraction solvents, including diethyl ether,

dichloroethane, and dichloromethane, showed that traps baited with
dichloromethane extract were the most consistent in trapping males.
In each of three nights about 52% of the dichloromethane extract traps
caught males, whereas only about 25% of the traps containing the other
extracts captured males. It was pointed out, however, that the ether
used was crude and that a redistilled ether extract that had been
stored at 0°-5° C for 14 weeks was still more active than an extract
with crude ether stored for only 6 weeks.

Although the cylinder trap had captured up to 38 moths in a trap in
a single night, it was judged to be an inefficient trap. In recent
studies, other trap designs have been used, including a sticky-tray
trap, similar to the Pherocon® trap (Zoecon Corp., Palo Alto, Calif.)[4],
and a water-basin trap (20 cm in diameter). The attractant-impregnated
paper is either suspended above the water, which contains some detergent, or floated on it. Data on the relative efficiency of these traps
were not available.

The preliminary data led the way to more intensified studies on the
identification of the pine caterpillar moth pheromone. The masses were
mobilized and thousands of moths were collected and snipped. The crude
pheromone extract was dissolved in acetone and cooled to -50° C. The
supernatant was chromatographed either on silica gel and Sephadex LH
20 columns or on silica gel thin-layer plates, and the fractions were
assayed for pheromone activity by field trapping tests. The most active
fractions were (1) the 10% ether in petroleum ether and the 100% ether
fraction from the silica gel column; (2) the 61-75 ml fraction from the
Sephadex column eluted with benzene:methanol (1:1); and (3) the R_f
0.60-1.75 band from the TLC plate developed with benzene:ether:acetic
acid (85:15:1). A more direct technique of working up the crude oil
extract was then found to give the active extract by EAG assay. The
crude oil was heated at 170° C in an oven under a stream of nitrogen
that subsequently passed through a dry-ice trap and finally into a
liquid nitrogen trap. Airborne collections also were being investigated by passing air over virgin female moths and bubbling it through
dichloromethane.

In preliminary chemical reaction tests, the EAG activity was not
destroyed when the crude oil was saponified and acetylated. Active
material from the columns or vaporization technique was collected from
a gas chromatographic column (Perkin-Elmer Model 900 gas chromatograph
equipped with a 3% neopentylglycol adipate on Chromosorb W, 80-100 mesh,
column @ 170° C) in 2-minute fractions. Strong EAG activity was found
in the 4-6 minutes fraction. Further study of compound obtained by
rinsing the exposed glandular surface of abdominal tips, precipitating
some impurities from acetone, and direct vaporization of the supernatant
at 170° C with nitrogen revealed the presence of several hydrocarbons
by computerized GLC-MS (MAT 311 mass spectrometer interfaced with a
Varian Aerograph gas chromatograph). One component, identified as
octadecane, was suspected of being involved in the pheromone system,
although no behavioral confirmatory tests had been conducted.

Concurrent with the chemical investigations of the pheromone at the
Peking Institute of Zoology are histological studies on the pine caterpillar sex pheromone gland. These studies have confirmed that the female gland, similar to other lepidopterous pheromone glands, is a

modified intersegmental membrane located between the eighth and ninth abdominal segments. The enormous surface area of the gland with its oval projections and spikes was observed with a scanning electron microscope.

The recent studies at Peking on the pine caterpillar moth have been carried out in cooperation with the Forest Pest Control Experimental Station of Kiangsi and the Institute of Applied Chemistry of Kirin. At the latter location, cooperative research on the identification of the pheromone structure is under way. The chemists in Kirin have carried out several tests on the crude extract from female pine caterpillar moths. Bromination resulted in a loss of activity; esterification some-times eliminated activity and sometimes it did not. Activity remained when the extract was treated with base or with phosphoric acid. The above results suggest that the compound is unsaturated but is not an ester or an epoxide. The esterification reaction is questionable, but an alcohol would explain the silica gel column chromatography data from Peking much more satisfactorily than the suggested hydrocarbon. Also, the support for unsaturation in the compound is not consistent with the suggested octadecane structure. There is a possibility that multiple components are involved and that each location has been working with a different component. Since the competence of the chemists at both in-stitutes is very good, coordination of the data and further research should result in a full identification of the pine caterpillar moth pheromone in the near future.

European corn borer Ostrinia (Pyrausta) nubilalis *(Hübner)* The chemists at the Kirin Institute of Applied Chemistry have also been concerned with the European corn borer pheromone since it is one of the major corn pests in China. A study of the pheromone of this insect has been very interesting in the United States because it was found that corn borers in Iowa were attracted to *cis*-11-tetradecenyl acetate (3% *trans*) (Klun *et al.*, 1973)[5], whereas corn borers found in New York are attracted to the opposite ratio, *trans*-11-tetradecenyl acetate (3% *cis*) (Kochansky *et al.*, 1975)[6]. The two populations co-occur in Pennsylvania without apparent crossmating (Carde *et al.*, 1975b)[7]. A study of European corn borer field responses in 10 countries in Europe showed that the *cis*-re-sponding populations were predominant and the *trans*-population was main-ly centered in Italy, one of the sources for broomcorn shipments to the United States during 1909-1914[8]. The chemists in Kirin synthesized *cis*-11-tetradecenyl acetate by the Wittig method, similar to the prepar-ation of this compound when it was reported as the sex attractant for the redbanded leafroller moth, *Argyrotaenia velutinana*[9]. The *trans* isomer was similarly synthesized by changing the solvent and base for the Wit-tig reaction. (An excellent review of the stereochemical control possible by controlled partial equilibration of the Wittig intermediates is found in ref. 10.) Field studies were conducted with these synthet-ics in cooperative work with entomologists at the Kirin Academy of Agricultural Science at Kungchuling.

The entomologists have studied the behavior of the European corn borer in the field and have had experience with using its pheromone for trapping male moths. They have found that the first-generation adults are found mainly in wheat; not many are found in millet. The second-generation adults are found mainly in soybean. It also was reported that the adults copulate around midnight, with all activity being over before dawn. In 1974, over 200,000 overwintering larvae were collected and the emerged females were used to obtain abdominal tip extract in dichloromethane, dichloroethane, and ethanol. The extracts were tested in the field with water-basin traps and impregnated cotton wicks. It was found that 20 female equivalents in dichloromethane attracted an average of 91.6 males per night, whereas the next best solvent, dichloroethane, attracted only 44.3 males per night. The crude extract was thought to have power to attract over a range of 50 m. Even though the power of attraction of crude extract was very good, the entomologists were not able to attract any males with the synthetic compounds. Isomeric purity of the Wittig-produced compounds was determined only by IR, so the amount of geometrical isomers present in each could be higher than desired for field use. Laboratory bioassays did show positive responses to the *cis* compound and to a mixture of *cis* and *trans* isomers--the *trans* compound alone was not tested. The chemists are now working on the crude extract to determine what compounds are produced by the females, so that the pheromone components can be used in disruption tests for insect control, as well as in tests in which the pheromone is combined with the *Beauveria* fungus, which is now being used for European corn borer control in Kirin.

An enthusiastic group at the Shanghai Institute of Organic Chemistry also is pursuing the European corn borer problem in cooperation with a pheromone research group at the Shanghai Institute of Entomology. They synthesized the *cis* and *trans* isomers by the more conventional alkyne intermediate route (Warthen, 1968; Schwarz and Waters, 1972)[11,12] as follows:

$$HO(CH_2)_{10}OH \xrightarrow[\text{2. dihydropyran}]{\text{1. HCl}} THP-O(CH_2)_{10}Cl \xrightarrow[\text{DMSO}]{LiC\equiv CH} THP-O(CH_2)_{10}C\equiv CH$$
$$I$$

$$I \xrightarrow[\text{EtBr}]{\text{BuLi}} THP-O(CH_2)_{10}C\equiv CCH_2CH_3 \xrightarrow[\text{2. HOAc/AcCl}]{\text{1. H}_2/\text{Lindlar}} CH_3\overset{O}{\overset{\|}{C}}O(CH_2)_{10}\overset{H}{\overset{|}{C}}=\overset{H}{\overset{|}{C}}CH_2CH_3$$
$$cis$$

$$\xrightarrow[\text{2. HOAc/AcCl}]{\text{1. Na/NH}_3} CH_3\overset{O}{\overset{\|}{C}}O(CH_2)_{10}C=\overset{H}{\overset{|}{C}}CH_2CH_3$$
$$\underset{H}{|}$$
$$trans$$

The isomers were purified by silver nitrate-silica gel column chromatography before use in the field. The isomers, or various combinations of them, were not attractive to European corn borer moths. These results, combined with those obtained in Kirin, indicate that the corn borer in China uses a different pheromone system than that found in other countries, possibly using an additional obligatory third component. Further research on the pheromone of this insect in China will be of much interest to many scientists around the world.

Pink bollworm Pectinophora gossypiella (*Saunders*)   The groups at the Shanghai Institute of Chemistry also have been vigorously pursuing the application of pheromones to control the pink bollworm, a major cotton pest.   These chemists first synthesized hexalure, *cis*-7-hexadecenyl acetate[13] (a parapheromone with much less attractant activity than the natural pheromone), and then synthesized the pheromone, gossyplure[14] (a 1:1 mixture of *cis*-7, *cis*-11- and *cis*-7, *trans*-11-hexadecadienyl acetates, IV and V).   The synthetic route was similar to that reported by other investigators[15,16] which represent only one of several routes reported for the synthesis of these compounds.[17,18,19]

$$CH_3(CH_2)_3C \equiv CH \xrightarrow[\triangle \; O]{\text{Grignard}} CH_3(CH_2)_3C \equiv C(CH_2)_2OH$$

$$\xrightarrow[\text{2. PBr}_3]{\text{1. H}_2, \text{ Lindlar}} CH_3(CH_2)_3 \overset{H}{C} = \overset{H}{C}(CH_2)_2Br \quad \text{II}$$

$$\xrightarrow[\text{2. PBr}_3]{\text{1. Na/NH}_3} CH_4(CH_2)_3 \overset{H}{C} = C(CH_2)_2Br \quad \text{III}$$

$$\text{II or III} \xrightarrow[\text{THPO(CH}_2)_6C \equiv CH]{\text{LiNH}_2/\text{NH}_3} THPO(CH_2)_6C \equiv C(CH_2)_2CH = CH(CH_2)_3CH_3$$

$$\xrightarrow[\text{2. HOAC/AcCl}]{\text{1. H}_2, \text{ Lindlar}}$$

$$cis,cis(\text{IV}) \text{ or } cis,trans(\text{V})$$

    The pheromone research group at the Shanghai Institute of Entomology is composed of entomologists and chemists and appears to be carrying out the most extensive experiments on using pheromones for insect control.   In addition to their work on the European corn borer, they have several years of experience with pink bollworm trapping in which females, crude extract, hexalure, and gossyplure were used.   With gossyplure now available from the Shanghai Institute of Organic Chemistry, they have conducted field trials for control of the pink bollworm by the mass trapping technique as well as by the disruption technique.
    Mass trapping.   Water-basin traps (20 cm) were baited with 50-11 pg of gossyplure in either polyethylene caps or polyethylene bags.   In 1974, traps were positioned throughout a 27-hectare cotton field at the rate of 30 traps per hectare.   Although the three generations of pink bollworms were subjected to six sprays of dichlorvos and dimethoate, 290,000 male pink bollworms were captured in the traps.   This number was estimated to be only 25% of the total number of males in the field. Over 1,000 females captured in blacklight traps each generation were checked for spermatophores.   A comparison with mating frequency in an untrapped cotton field showed no differences in the mating rates.   This experiment is continuing in 1975, but populations are obviously too large to control with the mass trapping technique.
    Attraction disruption   In 1974, the disruption technique was evaluated by permeating the air with gossyplure in 2-hectare test plots. Filter paper strings containing 20 pg of gossyplure were placed directly on cotton plants in every third square meter in one plot, and were put in polyethylene bags and distributed at the same rate in another

plot. Fresh strips were added to the "exposed paper" plot every 2 days; polyethylene bags were added to the "bag" plot every 7 days. Monitoring traps containing five female equivalents of pheromone extract were positioned in each plot to test for suppression of male attraction. The two methods of distributing gossyplure gave similar results, with an approximate 90% decrease in male catch over 50 days compared with catches in a check plot (about 1,000 males were trapped in the check plot). The results were encouraging, and in 1975 new experiments were being conducted to test various gossyplure load rates and filter paper densities. The basic research on disruption could rapidly lead to practical use of pheromones in the communes if the results continue to be positive. It was reported that field research on the pink bollworm moth had already involved 139 production teams in 44 production brigades in Hon-hu County of Hupei Province.

Oriental Fruit Moth Grapholitha molesta *Busck* Interest in utilizing the pheromone of the oriental fruit moth has resulted in its synthesis by at least two groups, one at the Peking Institute of Zoology and the other at the Department of Insecticide Assay at Tientsin. The pheromone, consisting of *cis*-8-dodecenyl acetate, 7% *trans*-8-dodecenyl acetate, and 300% dodecan-1-ol[20,21,22], was first synthesized by a conventional 10-step route using hexamethylphosphortriamide (HMPT) as a solvent in the coupling reaction[12]:

$$CH_3CH_2CH_2C\equiv CH + Br(CH_2)_7OTHP \xrightarrow[\text{HMPT}]{\text{BuLi}} CH_3CH_2CH_2C\equiv C(CH_2)_7OTHP$$

They have recently changed to an improved six-step synthesis better suited for large-scale production[23]. The first step uses readily available starting materials and avoids the handling of large amounts of lithium acetylide:

$$Cl(CH_2)_6I + CH_3(CH_2)_2C\equiv CH \xrightarrow[\text{NH}_3]{\text{NaNH}_2} Cl(CH_2)_6C\equiv C(CH_2)_2CH_3$$
$$VI$$

The next three reactions have been carried out by the Australians in a 91% overall yield (Holan and O'Keefe, 1973)[23], using sodium bis-(methoxyethoxy)aluminum hydride to reduce the acetylenic acid instead of lithium aluminum hydride:

$$VI \xrightarrow[\substack{\text{NaI} \\ \text{EtOH/H}_2\text{O}}]{\text{NaCN}} \xrightarrow[\text{H}_2\text{O}]{\text{NaOH}} \xrightarrow{\text{LiAlH}_4} HO(CH_2)_7C\equiv C(CH_2)_2CH_3$$

The final two steps involve acetylation and subsequent reduction with disiamylborane to the *cis* isomer.

Field tests are being conducted to determine the optimum *cis* per *trans* ratio for attraction of the oriental fruit moth in China. This is an important first step since oriental fruit moth pheromone responses are not the same in several other countries (G. Rothschild, Australia, personal communication).

Paddy Borer Tryporyza incertulas *Walker* The paddy borer remains one of the major pests of rice in China. There is much interest in identifying the pheromone of this insect, and several groups have initiated projects with this purpose. At the Shanghai Institute of Entomology, a method has been worked out for rearing the paddy borer on an artificial diet[24]. The pheromone research group has started preliminary work on the pheromone. Another group at the Kwangtung Institute of Entomology also is initiating a project on the identification of the paddy borer pheromone. They will collect specimens from the fields for this research since the insect is readily available at various times of the year. The chemistry will be carried out in cooperation with the Department of Chemistry, Changshan University, as well as the Shun te County Hormone Experiment Factory. Possibly chemists in the Department of Plant Protection, Kwangtung College of Agriculture and Forestry, who have been researching the application of the sterility technique for the control of the paddy borer will assist[25].

The above studies represent the major efforts on pheromones in China. Several other pest species were mentioned as possible target insects for pheromone studies, but apparently work has not yet started on them. There is much enthusiasm for the research on pheromones, and the absence of governmental and industrial roadblocks gives China the possibility of being the first country to use pheromones for insect control in annual large-scale pest management programs.

Insect Hormones

Sericulture We purchased some silk at the "friendship" store in Canton and wondered if it indeed had come from "hormone-treated" silkworms. We had just visited the Shun-te County Hormone Experiment Factory, where a juvenile hormone mimic (Sláma *et al.*, 1974)[26], or insect growth regulator (IGR), is synthesized and supplied to production brigades specializing in sericulture. A visit to one such production brigade allowed us to observe the actual use of an IGR that has given most brigades a 15% increase in silk production and in 1974 helped to give China the highest production of raw silk in its history. At the silkworm brigade, the young silkworm larvae were contained in shallow round baskets (about 1 m in diameter for 10,000 larvae) and fed cut-up mulberry leaves every 4 hours. The fifth instar larvae were transferred to rooms and dispensed in a dense 0.5-m row around the outer limits of the room. A solution of IGR then was sprayed from a conventional hand sprayer onto this mass of larvae. The resulting larvae were larger, spun larger cocoons, and thus produced more silk (Figures 9, 10 and 11).

The above procedure sounds, and even appeared to be, quite simple, but it is the result of a much basic research by many laboratories around the world on the development of IGR's for application on harmful, as well as useful, insects. Some of the questions that must be answered prior to the use of IGR's in sericulture include: Which chemicals are economically the best? How much IGR should be applied on the larvae for the best effect? When should the IGR be applied on the larvae?

FIGURE 9 Production of 20,000 silkworm larvae.

FIGURE 10 Spraying hormone on fifth instar silkworm larvae.

FIGURE 11 Showing the results of hormone-treated larvae. The resulting larvae were larger, spun larger cocoons, and thus produced more silk.

The identification[28] of the structure of a natural juvenile hormone
(VII, Table 4) from the *Hyalophora cecropia* moth opened the door for
a flurry of molecular modification activities. The driving force was
the anticipation of using IGR's to control insect pests by altering
the growth processes affected by juvenile hormone. Much of the effort
was directed toward lethal effects on the insects. However, some Japanese scientists (Akai *et al.*, 1971)[27] found that an injection of the
juvenile hormone VII into *Bombyx mori* larvae during the first half of
the fifth instar resulted in a prolonged larval life and a 30% increase
in the cocoon shell weight. The direct effects appeared to be increased RNA and fibroin synthesis. Further research[29] with the hormone
VII and an 11-propyl analog (VIII, Table 4)[30] showed that the latter
compound was much more active than the natural hormone. Oral or topical administration of 0.5 µg per larva of VIII (compared with 40 µg per
larva of VII) produced increases of 40% in silk protein. The increase

TABLE 4 IGR's that induce the development of
giant silkworm cocoons.

Country of Silkworm Assays	Compound (No. this chapter)	References Activity	References Chemistry
Japan	VII	28	27
	VIII	29	30
	IX	31,32	33
	X	31	32
	XI	—	34 Altosid® Zoecon Corp.
	XII	35,36	37
	XIII	36	37
	XIV	38	38
People's Republic of China	XV	39,40	41 R-20458 Stauffer Chem.
	XVI	39	42 USDA
	XVII	39	42 USDA JH-25
Soviet Union	XVIII	43	34 Altozar® Zoecon Corp.
	XIX	—	41,45

R = alkyl

R',R" = alkyl, alkoxy, carbalkoxy, NO$_2$, NH$_2$, CN, SO$_3$H, halo was not in the diameter but in the length of the silk filament. It also was important to find that there was no decrease in the number of reelable cocoons, in the quality of cocoon on the cocoon filament.

At the same time, similar studies in Japan with some methylenedioxy-phenyl derivatives (XII, Table 4) were being carried out. If a dosage of 10 µg per larva was applied topically, the larvae would continue to grow and subsequently undergo malformation and death without cocoon spinning. Lower dosages (1 µg per larva for XII and 5 µg per larva for XIII), however, were effective in prolonging the instar duration and producing up to 50% increases in weight. Applications on 5th instar larvae at 96 or 120 hours were the most effective, whereas application made at 168 hours resulted in malformation and death. Oral administration of 8 ppm during the 72nd-96th hour of the 5th instar also induced increases in cocoon-layer weight of 20%-35%.

More recently, a host of IGR's (Table 4 and many related analogs) have been tested in Japan and Russia for their effectiveness in producing giant cocoons. A number of the IGR's already had been developed for their effectiveness in controlling some insect pests. The Chinese have concentrated their efforts on several compounds developed in the United States. The most widely used IGR in China is a geranyl phenyl ether derivative (XV, Table 4) developed by Stauffer Chemical Co. (R-20458). Much research has been conducted on its potency and selectivity in insects[47] and on its mammalian metabolism and environmental degradation[48,49]. Compound XV was found to have good activity for affecting giant cocoons and is easy to synthesize. It has been synthesized in small amounts at the Peking Institute of Zoology and at the chemistry department at Chungshan University in Canton and in larger amounts at the Shanghai Institute of Organic Chemistry and at the Shun-te County Hormone Experiment Factory, Kwangtung Province. The Hormone Experiment Factory produced at least 10 kg of XV in 1974 in their single laboratory setup, using the following synthetic route:

In an adjacent laboratory at this factory, chemists are producing kilograms of N,N-bis(phosphonomethyl)glycine (Polaris®, CP 41845, Monsanto), which is used to treat sugarcane to improve juice purity and sucrose content.

The other IGR's under consideration for practical use in China are the methoxy- (XVI, Table 4) and ethoxy- (XVII) derivatives of XV. Many other analogs, such as those synthesized at the Peking Institute of Zoology (Table 5), were also prepared but were found to have little or no activity at dosages of 0.1-5 µg per larva. The chemists at the Shun-te County Hormone Experiment Factory prepared a series of derivatives in which the ethylphenyl group of XV and XVII were replaced with

phenyl or acetophenone[49]. The compounds, which apparently were not
very active in producing giant cocoons, were characterized by mass
spectra and NMR (60 and 250 mHz). The analyses revealed that the

TABLE 5 List of chemicals
showing little or no activity
with 5th instar silkworm larvae.

large-scale synthesis of XV gave only 30%-60% monoepoxidized product,
with the major impurity being the diepoxy derivative. The sericulture
industries use the 30%-60% product without further purification.

Cooperative research[39] involving the Kiangsu Institute of Sericulture, the Academy of Agriculture and Forestrial Sciences of China, the
Peking Institute of Zoology, and the Shanghai Institute of Organic
Chemistry was conducted on a bivoltine race of B. mori, autumn brood,
to determine the efficacy of IGR's XV, XVI, and XVII in increasing silk
production[39]. It was found that XV had high activity but that dosages
higher than 0.5 µg per larva resulted in a decreased number of spinning
larvae. Compound XVI gave the same 10%-15% increase in silk production
but could be given in dosages up to 5 µg per larva, with the optimum
at 2 µg per larva. This compound, therefore, has a greater margin of
safety than does XV. The optimum dosage, however, varies with the time

of application. With compound XVI, 5 µg per larva gave the best results if applied at 48 hours after 5th instar commenced feeding, whereas 4 µg was best at 60 hours, and 3 µg was best at 72 hours. Application
after 72 hours usually resulted in a larger number of nonspinning silkworms. The results at 72 hours were generally the most consistent and
showed the highest yields of silk.

The best time for spraying of IGR's also must be adjusted according
to climatic conditions[40]. In general, the silkworms are sprayed earlier in warm temperatures and later when the temperature is low. An
example of this is found in the variations observed in Shantung Province. In the central, western, and southern parts of this province,
the spring and mid-autumn silkworms are sprayed at 84-96 hours into
the 5th instar, whereas the summer and early-autumn silkworms are
sprayed at 72-84 hours into the 5th instar. At the Geotung Semi-
Island, the spring and mid-autumn silkworms are sprayed at 96-108 hours,
whereas the summer and early-autumn silkworms are sprayed at 72-84
hours in the 5th instar. It was also noted that the rearing temperature should not be lower than 23° C or many of the larvae will not
spin. Another variation is that more IGR is applied to larvae of the
bivoltine race than to those of the polyvoltine race.

In studies by the cooperative institutes on how the IGR's affect
the feeding habits of the larvae, it was found that the treated larvae
feed for 1-1.5 days longer than the control larvae but do not eat as
many leaves in the first 3 days after treatment. This physiological
response to the IGR treatment means that fewer mulberry leaves need
to be used for the IGR-treated larvae. In fact, an additional study
showed that increased quantities of silk, as well as reduced leaf consumption, were obtained when the amount of mulberry leaves was controlled, as opposed to allowing the larvae to eat freely.

The formulation of IGR for spraying also can affect its activity.
The compounds are usually formulated for the commune in a Tween® 80%
water emulsion containing 1%-4% ethanol. For practical use, it is
supplied in ampoules, whose contents are diluted in a hand sprayer so
that 4:1 of spray, containing 20 mg of XV or 50 mg of XVI or XVII, is
distributed on 40,000 larvae. A study at the Institute of Sericulture
of Shantung Province showed that several other types of emulsion were
superior[40]. They prepared 19 different emulsions of compound XV, which
was supplied by the Department of Chemistry, Chungshan University,
Kwangtung Province. One particularly good emulsion consisted of 12-24
g of concentrated oil of XV, 150 ml of emulsifying agent "0203" or
"0204," and 95% ethanol to give a total volume of 1,000 ml. Another
good emulsion was prepared with benzene:chlorobenzene (1:1) in place
of the ethanol. The emulsions were tested by spraying 0.15 ml per
larva at 96 hours into the 5th instar and evaluating the physiological
activity by the duration of the 5th instar and the weight of the cocoon
layer. Large-scale experiments showed that the application of the emulsions was convenient and effective in increasing silk production, and
it was observed that there were no changes in activity of the emulsion
after being stored at room temperature for 4-6 months.

In addition to the experiments on the use of IGR's in sericulture,
several groups are conducting basic research on what processes are

affected by the IGR's and why there is an increase in silk production. At the Peking Institute of Zoology, it was found that in addition to the physiological response of decreased food intake during the first 3 days after treatment with IGR the larvae also exhibited decreases in rate of silk protein synthesis, rate of body weight increase, and amylase and proteinase activities of digestive juices. These factors increased after the first 3 days and finally exceeded these activities in control larvae because of the lengthened larval stage of the treated larvae. Similar results were obtained at the Hormone Experiment Factory. The optimum timing of IGR application was found to coincide with the period when the natural juvenile hormone level had decreased to almost zero and ecdysone was just starting to build up. This was also at the time when silk protein was increasing tremendously. One effect of the IGR treatment appeared to be to decrease the buildup of plasmatocytes in the haemolymph that normally takes place in the middle of the 5th instar to destroy larval imaginal discs. Radiotracer studies showed that the DNA content in larvae after treatment initially increased above that of control larvae, but after 24 hours was lower, whereas the opposite effects were found with RNA. A similar effect was found with trans-aminase in the silk gland. The rate of enzyme increase in treated larvae initially was lower than that in control larvae, but after 1 day the enzyme was accumulating at a higher rate than in control larvae. This was given significance because of its role in converting glutamic acid in mulberry leaves to alanine, one of the major components of silk fibroin.

Ecdysone-related effects Ecdysones are widely distributed in insects and higher plants and are important in insect development. Early studies[50],[51],[52] described the effects of ecdysone and ecdysterone on insect protein synthesis (Figure 12). This suggested that they also might

FIGURE 12 Ecdysone
and related
phytoecdysones.

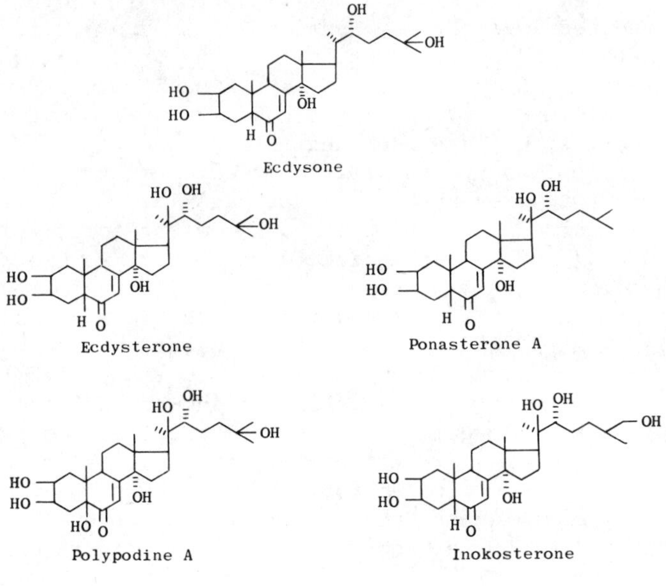

Ecdysone

Ecdysterone

Ponasterone A

Polypodine A

Inokosterone

affect fibroin synthesis in the posterior division of the silk gland of
B. mori. A preliminary study in Japan had indicated that phytoecdysones obtained from *Achyranthis radix* could cause a decrease in silk
formation[53]. Further investigations in Japan[54] showed that ecdysterone
is important in the promotion and maintenance of fibroin synthesis, but
its effect varies with the larval age (Shigematsu and Moriyama, 1970)[54].
Injection of feeding fifth instar larvae of various ages with ecdysterone revealed an inhibition of fibroin synthesis in 3-day-old larvae, and
a significant stimulation in 5-, 6-, and 7-day-old larvae. It was suggested that the effects are related to the amounts of ecdysterone already present in the larvae. Sufficient ecdysterone is obtained from
mulberry leaves (1 µg in 1 g of leaf powder[55]) in the first half of the
instar to maintain synthesis and negate effects of additional exogenous
supplies[55]. The finding that fibroin synthesis was much higher in larvae fed on diets containing ecdysterone than in larvae fed on ecdyster-
one-free diets was significant for the development of silk factories
that utilize silkworms reared on a cheap artificial diet instead of
mulberry leaves.

Further studies in Japan[56] showed that inokosterone (Figure 1),
another phytoecdysone of mulberry leaves[55], also is an important stimulus for fibroin synthesis. Silk formation in larvae fed on a diet containing 5 µg of ecdysterone or 10 µg of inokosterone per 1 gram of dry
diet was found to be lower than in controls, whereas silk formation
was greatly increased in larvae fed a diet containing 40 µg of inokosterone. Ponasterone A (Figure 1), which is not found in mulberry leaves,
was quite toxic to 4th instar larvae and inhibited the growth of silk
glands during the 5th instar. The effects, then, were found to be
dependent on the type, quantity, and time of application of ecdysone
used.

Research also has been conducted in China on the effects of phytoecdysones on 5th instar silkworm larvae[39]. Phytoecdysone extracted
from the roots of *Achyranthes bidentata* and ponasterone A were used to
shorten the duration of the last larval instar. If administered in the
food of the larvae, the ecdysones effected a decrease in silk production. If they were sprayed in doses of 2-5 g per larva after 10% of
the larvae had reached maturity, the instar was shortened by 12 hours
without any decrease in silk production or in quality of the cocoons.
At the Shanghai Institute of Organic Chemistry, the phytoecdysones
were extracted also from *Ajuga ciliata*, *A. decumbens*, and *A. nipponisis*.
The best results were obtained with *A. decumbens*. The phytoecdysones
are present in these plants at 0.1% of the dry weight as a 3:2 mixture
of ecdysterone and polypodine B. The purpose of this research is to decrease the amount of mulberry leaves and the labor involved, synchronize
the spinning larvae, and use the phytoecdysones in conjunction with an
IGR for increased silk production with a minimum quantity of leaves
and labor.

Juvenile hormone effects in termites Several species of termites are
major pests in China and are the focus of studies on their biology and
ecology. One research group at the Shanghai Institute of Entomology
has been studying the effects of IGR's on caste differentiation of

132

Reticulitermes flaviceps (Kollar) a major pest in buildings. The
ultimate goal is to control termites by upsetting the balance among
castes and disrupting the normal social life of a colony.

It has been known for many years that caste determination is under
hormone control[57], and that at low doses the juvenile hormone is responsible for inhibiting transformation into the final reproductive form.
In large doses, however, juvenile hormone can induce soldier development from pseudergates. This was determined first by transplanting
the large corpora allata of primary and secondary reproductives or of
nymphs just before the alate moult into pseudergates of *Kalotermes
flavicollis* and observing the great number of transformations to presoldiers[58]. This same phenomenon was observed by treating the pseudergates with juvenile hormone or related IGR's[59,60,61]. For example[61],
pseudergates of *Reticulitermes lucifugus santonensis* were exposed to
filter papers treated with 0.2 ml of acetone solutions of varying concentrations of ethyl 10-epoxy-3,7,11-trimethyl-2,6-dodecadienoate (an
IGR supplied by Zoecon Corp.). Solutions containing 0.05% IGR induced
the development of many pre-soldiers, whereas 0.5% solutions resulted
in a high mortality.

The Chinese have conducted similar studies on the related termite
R. flaviceps[62]. They used several geranyl phenyl ether derivatives
(including XI, XVIII, and the ethyl thiolate analog of XI). The experimental procedure was similar to that described above, with 0.2 ml
of 1%, 0.5%, 0.05%, or 0.005% acetone solution of IGR placed on a piece
of filter paper. The geranyl phenyl ether derivatives were without
effect on the test group of 50 pseudergates. The three Zoecon compounds mentioned above were quite active in inducing the pseudergates
to differentiate to presoldiers, which are distinguished from soldiers
by their white heads and mandibles. Compound XI showed the most activity, with significant effects produced by the 0.005% solution. High
mortality was obtained with the 0.5% solution of XI. The other two
Zoecon compounds were effective at concentrations down to 0.05%. The
presoldiers appeared 13 days after treatment, but the highest frequen-
cies were between 16 and 19 days. Since soldiers in a colony have a
suppressive effect on the production of more soldiers[63], an experiment
was conducted in which pseudergates were treated with XVIII in the
presence of a soldier. The results showed that the presence of a
soldier does not suppress the effect of the IGR on pseudergates. Recent studies at the Shanghai Institute of Entomology have indicated
that IGR's may be even more active if applied on the nymph forms.

A colony of lower termites, such as *R. flaviceps*, generally is composed of larvae, nymphs, pseudergates, presoldiers, soldiers, and
reproductives. Every individual possesses the genetic patterns to
differentiate into all the possible forms. The intercaste equilibria
are influenced by a number of environmental factors affecting the whole
population, by pheromones operating within the colony, and by the timing and intensity of the juvenile and moulting hormones. A colony of
R. flaviceps normally has a large number of pseudergates and a few
soldiers. The application of IGR creates an imbalance in the colony
by effecting the formation of a large excess of soldiers. The soldiers
will presumably emit pheromones to repress the formation of more

presoldiers, but the continued presence of an IGR can evidently overcome this natural feedback system and possibly throw the colony into
complete disarray. Even if this is shown to be true in laboratory
situations, however, the practical use of this technique will be extremely difficult for termite colonies in their natural habitat.

Insect Trapping

Light traps On an all-night train ride from Changsha to Canton, we
observed the purplish glows from scattered points throughout the fields.
We had been told that there were about 1 million blacklight traps
operating in this province of 4,000 km^2, and the amazing density of
traps seen from the train helped to confirm that report (Figure 13).
Many fields had electric lines strung through the middle of the fields,
leading from one trap to the next. The light traps were used for
insect forecasting and for insect control. The forecasting traps
normally consist of a 200-W incandescent light operating above a jar
containing potassium cyanide. A variety of blacklight fluorescent-
lamp traps were seen, consisting of several kinds of electric grid
traps[64] and several types of baffle traps[65] with water basins. Although the electric grid trap in the United States has generally been
considered to have a greater collection efficiency than other types[66],
the Chinese have not utilized them very much because of operational
difficulties. At many places, the water-basin blacklight trap is seen
as a permanent cement structure in the fields. The water, which is
located beneath the large baffles, is covered with a thin film of
kerosene and DDT, or with a film obtained from the sticky water drained
from rice during cooking. The nutritious rice water containing the
protein-rich trapped insects is then fed to the fish.

In Hunan Province, the blacklight traps are dispensed at the rate of
one trap for every 3-4 hectares; in Kiangsu Province, one trap for 1 or
2 hectares; and in Kwangtung Province, one trap for 2 or 3 hectares.

FIGURE 13 A blacklight trap
typical of those seen in some
countryside areas.

This represents a lot of traps, but the spacing is more reasonable than that usually found in other countries. Since the range of a blacklight is only several hundred feet, it has been calculated[67] for the tobacco budworm, *Heliothis virescens*, and the cabbage looper, *Trichoplusia ni*, that the capture potential is much greater at a spacing of two traps per hectare than at one trap per 2 hectares or at less dense spacings. It was pointed out[67] that the spacings used by most investigators have not been close enough for the control of phototactic insects. Nevertheless, a density of only one trap per 86 hectares was used by the U.S. Department of Agriculture[68] in a large blacklight-trapping experiment on St. Croix, an island in the Caribbean Sea (Cantelo *et al.*, 1974)[68]. A total of 250 blacklight traps were used on this island (218 km^2) for 43 months, and the effect on 25 insect species was measured. The conclusion was that blacklight traps had a significant effect on the populations and that they have the potential to be a control device if the trap area is sufficiently large. It also was observed that after the traps were removed, the populations resurged to levels much higher than the initial ones, presumably because the predators and parasites also had been reduced in numbers during the trapping years. In view of the fact that the traps in Hunan Province are used at closer spacings and are continued every year, there is a good possibility that insect control is effected for some species. The chances for control also are increased by continually servicing the traps and collecting egg masses in the vicinity of the traps.

The use of blacklight traps as a cultural control technique was evident in many of the areas of China that we visited. Their use, however, was being questioned by some researchers in the Kwangtung Province because of the large number of beneficial insects that were being trapped. With a reduction in the populations of pest insects, the number of beneficial insects would decline anyway, and so it is still possible to have the competing populations stabilize at a low level as long as the traps are continually used as the suppressing factor.

In some of the cotton-growing regions, blacklights have been used on a large scale since 1969 for the control of *Heliothis armigera* (Hübner). They have been found to decrease population densities in the field, but the trapping efficiency is not good compared with that obtained in trapping other insects. Therefore, research has been conducted at the Peking Institute of Zoology on some of the basic aspects of the phototactic behavior of this insect. The response of a number of insect species to 13 monochromatic lights ranging from 333 nm to 656 nm was observed[69]. It was found that 333 nm was the most effective with *H. armigera* and *Heliothis assulta* Guenee, whereas 375 nm was the most effective for armyworm adults. The blacklight normally used in the field emits a continuous spectrum from 300 nm to 500 nm, but the peak of the energy distribution curve is at 365 nm. The radiant intensity at 333 nm is only one-half that at 365 nm. Since the 333-nm wavelength emission is 2.5-3.4 times as effective as that at 365 nm, it was suggested that the blacklight should be modified for the *Heliothis* species.

Another experiment investigated the effects of combining visible light with ultraviolet light. The purpose was to determine if decreased

phototactic responses are caused by a lack of attraction or by the
presence of disturbing factors. The test utilized a box emitting
ultraviolet light from the other side. The radiation intensity of the
visible light was slightly lower than that of the ultraviolet light.
It was found that armyworms were captured in the greatest numbers by
a combination of 350-nm and 405-nm lights, but that combinations of
350 nm and visible lights above 436 nm had a detrimental effect on the
catches. The *Heliothis* adults were captured in larger numbers by combination of 350 nm and a range of visible lights of short wavelengths,
and only visible lights of long wavelengths were detrimental to the
catches. These differences were suggested as explanation for the fact
that blacklight traps captured more armyworms, whereas mercury vapor
gaseous discharge lamps captured more *Heliothis* adults.

The combination effect also was studied at the Nantung Institute of
Agriculture, Kiangsu Province, where they found in 1973 that a 20-W
blacklight trap caught 11,035 insects, while a combination trap using a
20-W blacklight lamp and a 200-W tungsten filament lamp caught 63,097
destructive insects (Nantung Institute of Agriculture, 1975)[70]. In 1974,
the blacklight trap caught 14,956 destructive insects, a combination
trap using a 20-W blacklight lamp and a 20-W white fluorescent lamp
caught 40,573 insects, and a combination trap using a 20-W blacklight
lamp and a 200-W tungsten filament lamp caught 38,940 destructive insects.

The results of the experiment with combinations of blacklight with
various wavelengths can give positive or negative effects on "attractancy." To further investigate this, scientists at the Peking Institute
of Zoology studied the changes in dark-adapted insect eyes when exposed
to various types of light stimuli. Normally, the compound eye of a
nocturnal moth is protected in strong light by a screening pigment
that decreases the light energy passing through. The dark-adapted
eye is 1,000 times more sensitive to light than the light-adapted eye[71].
Heliothis moths were placed in a closed box for 1 hour to allow the eyes
to become dark-adapted, and then they were exposed for 15 minutes to
one of five different ultraviolet lights. The head was then quickly
cut off, fixed, and cut into sections. A light microscope was used to
observe the inward migration of the screening pigment from the distal
portion of the eye. A transparent zone observed between the distal end
of the pigment and the crysalline cone was found to be the widest in
eyes illuminated by 333-nm light and was the narrowest in eyes illuminated by 365-nm light.

The changes in dark-adapted eyes to various monochromatic lights
also were studied by using a photomultiplier to measure reflected light
from the compound eye. A dark-adapted eye reflects about 70 times more
light than a light-adapted eye[71]. If the eyes were illuminated with an
ultraviolet light, the eyes quickly decreased in their ability to reflect light as a result of the rapid migration of the screening pigment.
A green or red light caused the reflecting properties of the eye to
change much more slowly.

The above results showed that a 333-nm light caused the greatest
changes in the dark-adapted eye, indicating that the eye could least
endure this light. The Chinese scientists speculated that a moth does

not fly into the light because of "attractancy" or because the eye is
sensitive to a particular wavelength, but that it flies into the light
trap because the dark-adapted eye cannot endure the light. The moths
are stunned by the light (like a rabbit caught in the headlights of
a car) and blindly continue their flight into the trap. The effectiveness of this blinding effect is dependent on the amount of background
light, so would be more effective on nights without much moonlight. An
explanation for the results of the tests utilizing a combination of
two different lights is that some combinations increase the dazzling
effect, whereas some visible lights simply increase the background
light and result in a decrease in effectiveness. A theoretical model
of how the insect computes the difference between light input from the
lamp and from the background is being developed and should be very interesting from an academic standpoint and in its value to the development of better blacklight traps.

Bait pails The use of bait pails is another common practice in China
(Figure 14). Most of the places use the common 6:3:1 mixture of sugar:
vinegar:wine mixed in nine parts of water. In Hunan Province, however,
some of the pails were baited with a local preparation. The fruit of
an ornamental tree, which is used for medicinal purposes in the neighboring Szechwan Province, is pounded into a pulp and allowed to ferment.
The addition of a small amount of water completes the recipe for the
bait pails. Although the pails are used in conjunction with the pest
control programs on different crops for various species, one of their
uses in Hunan Province was for control of the early-season black cutworm adults. In a vegetable commune outside of Shanghai, the pails
were used in a density of 60 pails per hectare in some of the vegetable
plots. In Kirin Province, the pails were used in an experiment to determine the flight dispersal range of the armyworm *Mythimna separata*.
About 600,000 dye-labeled moths were released in three central provinces
(Shantung, Kiangsu, and Anhwei). Three hundred moth-trapping stations,
each consisting of several dozen bait-pail traps, were set up in the
northern provinces of Kirin and Liaoning. Of the 2,600,000 moths collected, 6 were found to be labeled--several of which had traveled the
maximum distance. The experiment showed that the armyworm is capable
of long-distance flights and that the bait pails are capable of trapping
a huge number of insects.

FIGURE 14 A bait pail typical
of those used for insect control in some vegetable-producing areas.

REFERENCES

1. Section of Insect Pheromone, Peking Institute of Zoology. 1973. Preliminary studies on the sex attractant of the pine caterpillar moth, *Dendrolimus punctatus* Walker. Acta Entomol. Sinica *16*:94-96.
2. Roelofs, W. and A. Comeau. 1971. Sex pheromone preception: electroantennogram responses of the redbanded leafroller moth. J. Insect Physiol. *17*:1969-1982.
3. Roelofs, W. and A. Comeau. 1971. Sex attractants in lepidoptera. *In* "Chemical Releasers in Insects," Vol. III, Tahori, A. S. (ed.), Gordon and Breach, New York, N.Y. 91-114.
4. Bode, W. M., D. Asquith, and J. P. Tette. 1973. Sex attractants and traps for tufted apple budmoth and redbanded leafroller males. J. Econ. Entomol. *66*:1129-1130.
5. Klun, J. A., O. L. Chapman, K. C. Mattes, P. W. Wojtkowski, M. Beroza, and P. E. Sonnet. 1973. Insect sex pheromones: minor amount of opposite geometrical isomer critical to attraction. Science *181*:661-663.
6. Kochansky, J., R. T. Cardé, J. Liebherr and W. L. Roelofs. 1975. Sex pheromone of the European corn borer, *Ostrinia nubilalis* (Lepidoptera: Pyralidae), in New York. J. Chem. Ecol. *1*:225-231.
7. Cardé, R. T., J. Kochansky, J. F. Stimmel, A. G. Wheeler, Jr., and W. L. Roelofs. 1975. Sex pheromones of the European corn borer: *cis* and *trans* responding males in Pennsylvania. Environ. Entomol. *4*:413-414.
8. Klun, J. A. and Cooperators. 1975. Insect sex pheromones: Intraspecific pheromonal variability of *Ostrinia nubilalis* in North America and Europe. Environ. Entomol. *4*:891.
9. Roelofs, W. L. and H. Arn. 1968. Sex attractant of the redbanded leafroller moth. Nature *219*:513.
10. Anderson, R. J. and C. A. Henrick. 1975. Stereochemical control in Wittig olefin synthesis. Preparation of the pink bollworm sex pheromone mixture, Gossyplure. J. A. C. S. *97*:4327-4334.
11. Warthen, D. 1968. Synthesis of *cis*-9-tetradecen-1-ol acetate, the sex pheromone of the fall armyworm. J. Med. Chem. *11*:371-373.
12. Schwarz, M. and R. M. Waters. 1972. Insect sex attractants; XII. An efficient procedure for the preparation of unsaturated alcohols and acetates. Synthesis *10*:567-568.
13. Keller, J. C., L. W. Sheets, N. Green and M. Jacobson. 1969. *cis*-7-Hexadecen-1-ol Acetate (Hexalure), A synthetic sex attractant for pink bollworm males. J. Econ. Entomol. *62*:1520-1521.
14. Hummel, H. E., L. K. Gaston, H. H. Shorey, R. S. Kaae, K. J. Byrne, and R. M. Silverstein. 1973. Clarification of the chemical status of the pink bollworm sex pheromone. Science *181*:873-875.
15. Vick, K. W., H. C. F. Su, L. L. Sower, P. G. Mahany, and P. C. Drummond. 1974. (Z,E)-7,11-Hexadecadien-1-ol Acetate: the sex pheromone of the angoumois grain moth, *Sitotroga cerealella*. Experientia *30*:17-18.
16. Su, H. C. F. and P. G. Mahany. 1974. Synthesis of the sex pheromone of the female angoumois grain moth and its geometric isomers. 1974. J. Econ. Entolmol. *67*:319-321.

138

17. Mori, K., M. Tominaga and M. Matsui. 1975. Stereoselective synthesis of the pink bollworm sex pheromone, (Z,Z)-7,11-hexadecadienyl acetate and its (Z,E)-isomer. Tetrahedron *31*:1846-1848.
18. Bierl, B. A., M. Beroza, R. T. Staten, P. E. Sonnet, and V. E. Adler. 1974. The pink bollworm sex attractant. J. Econ. Entomol. *67*:211-216.
19. Sonnet, P. E. 1974. A practical synthesis of the sex pheromone of the pink bollworm. J. Org. Chem. *39*:3793-3794.
20. Roelofs, W., A. Comeau and R. Selle. 1969. Sex pheromone of the Oriental fruit moth. Nature *224*:723.
21. Roelofs, W. and R. Cardé. 1974. Oriental fruit moth and lesser appleworm attractant mixtures refined. Environ. Entomol. *3*:586-588.
22. Cardé, R., T. Baker and W. Roelofs. 1975. Behavioural role of individual components of a multichemical attractant system in the Oriental fruit moth. Nature *253*:348-349.
23. Holan, G. and D. F. O'Keefe. 1973. An improved synthesis of insect sex attractant: *cis*-8-Dodecen-1-ol Acetate. Tetrahedron, 673-674.
24. Research Group on Resistance to Insecticides, Shanghai Institute of Entomology. 1975. Rearing of paddy borer (*Tryporyza incertulas* Walker) on artificial diet under aseptic condition. Acta Entomol. Sinica *18*:128-132.
25. The Teaching and Research Group of Entomology and Pesticides, Department of Plant Protection, Kwangtung College of Agriculture and Forestry. 1974. A preliminary study on the application of the sterility technique for the eradication of the paddy borer, *Tryporyza incertulas*. Acta Entomol. Sinica *17*:135-147.
26. Sláma, K., M. Romaňuk and F. Šorm. 1974. Insect hormones and bioanalogues. Springer-Verlag/Wien, New York. 447 p.; Staal, G. B. 1975. Insect growth regulators with juvenile hormone activity. Ann. Rev. Entomol. *20*:417-460.
27. Akai, H., K. Kiguchi and K. Mori. 1971. Increased accumulation of silk protein accompanying JH [juvenile hormone] -induced prolongation of larval life in *Bombyx mori*. Appl. Entomol. Zool. *6*:218-220.
28. Röller, H., K. M. Dahm, C. C. Sweeley, and B. M. Trost. 1967. Die Struktur des Juvenilhormons. Angew. Chem. *79*:190-191.
29. Nihmura, M., S. Aomori, K. Mori and M. Matsui. 1972. Utilization of synthetic compounds with juvenile hormone activity for the silkworm rearing. Agr. Biol. Chem. *36*:889-892.
30. Mori, K., T. Mitsui, J. Fukami and T. Ohtaki. 1971. Synthesis of compounds with juvenile hormone activity. Part VII. A convenient non-stereoselective synthesis of the C_{18}-*Cecropia* juvenile hormone and its analogues; effect of the terminal alkyl substituents on biological activity. Agr. Biol. Chem. *35*:1116-1127.
31. Nihmura, M., S. Aomori, Y. Ozawa, K. Mori and M. Matsui. 1974. Effect of novel analogs of the juvenile hormone. I. Effect of juvenile hormone analogs with different alkyl substituents at C-7 and C-11 on the silkworm, *Bombyx mori* (Lepidoptera, Bombycidae). Appl. Entomol. Zool. *9*:34-40.

32. Ito, T., Y. Mizuta, K. Takamiya, S. Ueda, R. Kimura, T. Higuchi and S. Takahashi. 1975. Growth, development, and cocoon production in artificial diet-rearing of the silkworm, *Bombyx mori*, with or without application of a synthetic juvenile hormone analog. Nippon Nogei Kagaku Kaishi *49*:39-48.

33. Ozawa, Y., K. Mori, and M. Matsui. 1973. Synthesis of compounds with juvenile hormone activity. XV. Synthesis and biological activity of analogs of the Cecropia juvenile hormone with different alkyl substituents at C-7 and C-11. Agr. Biol. Chem. *37*:2373-2378.

34. Henrick, C. A., G. B. Staal, and J. B. Siddall. 1973. Alkyl 3,7, 11-trimethyl-2,4-dodecadienoates, a new class of potent insect growth regulators with juvenile hormone activity. J. Agr. Food Chem. *21*:354-359.

35. Chang, C.-F., S. Murakoshi, and S. Tamura. 1972. Giant cocoon formation in the silkworm, *Bombyx mori* L., topically treated with methylenedioxyphenyl derivatives. Agr. Biol. Chem. *36*:692-694.

36. Murakoshi, S., C.-F. Chiang and S. Tamura. 1972. Increase in silk production by the silkworm, *Bombyx mori* L., due to oral administration of a juvenile hormone analog. Agr. Biol. Chem. *36*:695-696.

37. Chang, C.-F., and S. Tamura. 1971. Synthesis of several 3,4-methylenedioxyphenyl derivatives as inhibitors for metamorphosis of silkworm larvae. Agr. Biol. Chem. *35*:1307-1309.

38. Tamura, S. and S. Cho. 1974. Aromatic ether derivatives. Japan Kokai 7,488,832, Aug. 24; C. A. *83*:43023r (1975).

39. Laboratory of Physiology and Pathology, Kiangsu Institute of Sericulture, Section of Sericultural Service, Academy of Agricultural and Forestrial Sciences of China; Laboratory of Insecticide and Toxicology, Peking Institute of Zoology, Academia Sinica; First Research Laboratory, Shanghai Institute of Organic Chemistry, Academia Sinica. 1974. Studies on the application of insect hormone analogues to improve silk production in *Bombyx mori* L. Acta Entomol. Sinica *17*:290-302.

40. Institute of Sericulture of Shantung Province. 1975. Experiments on the application of juvenile hormone analogue "734-II" emulsions to increase silk production of *Bombyx mori* L. Acta Entomol. Sinica *18*:266-272.

41. Pallos, F. M., J. J. Menn, P. E. Letchworth, and J. B. Miaullis. 1971. Synthetic mimics of insect juvenile hormone. Nature *232*: 486-487.

42. Sarmiento, R., T. P. McGovern, M. Beroza, G. D. Mills, Jr., and R. E. Redfern. 1973. Insect juvenile hormones: highly potent synthetic mimics. Science *179*:1342-1343.

43. Grigoryan, E. G., S. M. Sarkisyan, and G. Kh. Azaryan. 1973. Changes in the development and productivity of the silkworm under the effect of a juvenile hormone analog. Biol. Zh. Arm. *26*:26-51; C. A. *81*, 9994g (1974).

44. Samokhvalova, A. A. Drabkina, Z. A. Skachkova. 1974. Effects of a synthetic analog of a juvenile hormone on the silkworm larvae. Zool. Zh. *53*:1655-60; C. A. *82*, 134000b (1975).

45. Redfern, R. E., T. P. McGovern, R. Sarmiento and M. Beroza. 1971. Juvenile hormone activity of mixed ethers containing a phenyl and a terpenoid moiety applied topically to the large milkweed bug and the yellow mealworm. J. Econ. Entomol. *64*:374-376.

46. Menn, J. J. and M. Beroza. 1972. Insect Juvenile Hormones - Chemistry and Action. Academic Press, New York and London, 341 pp.

47. Gill, S. J., B. D. Hammock and J. E. Casida. 1974. Mammalian metabolism and environmental degradation of the juvenoid 1-(4'-ethylphenoxy)-3,7-dimethyl-6,7-epoxy-*trans*-2-octene and related compounds. J. Agr. Food Chem. *22*:386-395.

48. Hoffman, L. J., J. H. Ross, and J. J. Menn. 1973. Metabolism of 1-(4'-ethylphenoxy)-6,7-epoxy-3,7-dimethyl-2-octene (R20458) in the rat. J. Agr. Food Chem. *21*:156-163.

49. Bowers, W. 1969. Juvenile hormone: activity of aromatic terpenoid ethers. Science *164*:323-325.

50. Karlson, P. and C. E. Sekeris. 1962. Zum Tyrosinstoffwechsels der Insekten - IX. Kontrolle des Tyrosinstoffwechsels durch Ecdyson. Biochim. Biophys. Acta *63*:489-495.

51. Neufeld, G. J., J. A. Thompson and D. H. S. Horn. 1968. Short-term effects of crustecdysane (20-hydroxyecdysone) on protein and RNA synthesis in third instar larvae of *Calliphora*. J. Insect Physiol. *14*:709-804.

52. Arking, R. and E. Shaaya. 1969. Effect of ecdysone on protein synthesis in the larval fat body in *Calliphora*. J. Insect Physiol. *15*:207-296.

53. Ito, T., J. Koizumi, H. Yanagawa, M. Haroda, and A. Muroga. 1968. Acceleration of maturation of larvae of the silkworm by feeding of insect-molting hormone from the plants. Tech. Bull. Seric. Exp. Sta. *92*:21-40.

54. Shigematsu, H. and H. Moriyama. 1970. Effect of ecdysterone on fibroin synthesis in the posterior division of the silk gland of the silkworm, *Bombyx mori*. J. Insect Physiol. *16*:2015-2022.

55. Takemoto, T., S. Ogawa, N. Nishimoto, H. Hirayama, and S. Taniguchi. 1967. Isolation of insect-moulting hormones from mulberry leaves. Yakugaku Zasshi *87*:748.

56. Shigematsu, H., H. Moriyama, and N. Arai. 1974. Growth and silk formation of silkworm larvae influenced by phytoecdysones. J. Insect Physiol. *20*:867-875.

57. Lüscher, M. 1960. Hormonal control of caste differentiation in termites. Ann. N.Y. Acad. Sci. *89*:549-563.

58. Lüscher, M. 1969. Die Bedeutung des Juvenilhormons fur die Differenzierung der Soldaten bei der Termite *Kalotermes flavicallis*. Proc. VI Congr. IUSSI, Bern, 165-170.

59. Lüscher, M. and A. Springhetti. 1960. Untersuchungen uber die Bedeutung der Corpora Allata fur die Differenzierung der Kasten bei der Termite *Kalotermes flavicallis*. J. Insect Physiol. *5*: 192-212.

60. Hrdý, L. and J. Krecek. 1971. Development of superfluous soldiers induced by juvenile hormone analogues in the termite, *Reticulitermes lucifugus santonensis*. Insectes Sociaux *19*:105-109.

61. Hrdý, I. 1972. Der Einfluss von Zwei Juvenilhormonanalogen auf die Differenzierung der Soldaten bei *Reticulitermes lucifugus santonensis*. Z. ang. Entomol. *72*:129-134.
62. Chu, H.-H., T.-D. Tai, T.-F. Chen, and M.-W. King. 1974. Induction of soldier differentiation in the termite, *Reticulitermes flaviceps* Oshima with juvenile hormone analogues. Acta Entomol. Sinica *17*:161-165.
63. Light, S. F. and F. M. Weesner. 1955. The production and replacement of soldiers in incipient colonies of *Reticulitermes hesperus*. Insectes Sociaux *2*:347-354.
64. Kishaba, A. N., W. W. Wolf, H. H. Toba, A. F. Howland, and T. Gibson. 1970. Light and synthetic pheromone as attractants for male cabbage loopers. J. Econ. Entomol. *63*:1417-20.
65. Harding, W. C., Jr., J. G. Hartsock, G. G. Rohwer. 1966. Blacklight trap standards for general insect surveys. Bull. Entomol. Soc. Amer. *12*:31-32.
66. Goodenough, J. L. and J. W. Snow. 1973. Increased collection of tobacco budworm by electric grid traps as compared with blacklight and sticky traps. J. Econ. Entomol. *66*:450-453.
67. Hartstack, A. W., Jr., J. P. Hollingsworth, R. L. Ridgway, and H. H. Hunt. 1971. Determination of trap spacings required to control an insect population. J. Econ. Entomol. *64*:1090-1100.
68. Cantelo, W. W., J. L. Goodenough, A. H. Baumhover, J. S. Smith, Jr., J. M. Stanley, and T. J. Henneberry. 1974. Mass trapping with blacklight: effects on isolated populations of insects. Environ. Entomol. *3*:389-395.
69. Ting, Y.-C., T.-T. Kao and D.-M. Li. 1974. Study on the phototaxic behaviour of Nocturnal Moths. The response of *Heliothis armigera* (Hübner) and *Heliothis assulta* Guenee to different monochromatic light. Acta Entomol. Sinica *17*:307-317.
70. Nantung Institute of Agriculture, Kiangsu Province. 1975. Effects of light traps equipped with two lamps on capture of insects. Acta Entomol. Sinica *18*:289-294.
71. Agee, H. R. 1973. Spectral sensitivity of the compound eyes of field-collected adult bollworms. Ann. Entomol. Soc. Am. *66*: 613-615.

E. INTEGRATED INSECT CONTROL

General

Insect control in China in the old days was traditionally by burning, handpicking, and growing of native varieties that had some resistance to pests. After the revolution, modern scientific research and control methods were initiated. Plant protection now has high priority in China, and integrated control has been popularized and is increasingly practiced by many communes. Integrated control matches with the philosophy of Chairman Mao, who believes man can conquer nature and stresses manipulation of the environment to suppress pest attack. The Chinese are working to apply pest control in context with the

total agricultural production scheme, all activities being directed
toward ensuring a good harvest.

The application of integrated insect-control technology to agricultural pest problems in China far exceeded expectations of the delegation. The awareness in China of international progress in the
development of integrated control, particularly, it seemed, in the
United States, was excellent. Strong interest was evident throughout
the study period at all levels in the areas visited, starting the
first morning at the Peking Institute of Zoology with a general lecture on this topic to applied efforts of technical workers on production brigades in the communes visited later. The delegation was
greatly impressed with the dedication to the development of integrated
control in China.

Clearly, the Chinese have progressed beyond levels attained in the
United States both in widespread enthusiasm for integrated control and,
in many respects, in the application of the ecological principles
fundamental to its development. The only reservations we heard about
its utility were expressed at a vegetable-producing commune. However,
even at this commune, sophisticated forecasting services negated all
insecticide applications but those considered essential on the basis
of samples of insect populations. The development and application of
cultural control methods were exemplary in many instances. The terms
"pest management" or "integrated pest management" were virtually never
used by our Chinese colleagues; the term used was "integrated control."

The value of insecticides in crop protection is recognized throughout China; indeed, most of their serious pests are primarily controlled
with chemicals. Many of the problems associated with heavy dependence
on insecticides so familiar to us in the United States are becoming
increasingly apparent in China. These include depletion of beneficial
insect populations, insecticide resistance, increased importance of
secondary pest problems, crop residues, and environmental contamination.
To combat these increasingly severe problems, some of which could become limiting factors in crop production, nationwide attention is being
paid to research on and development of integrated control technology.
Most of the techniques in use or under development are suppressive
tactics alternative to the use of chemical insecticides. Yet, it is
recognized that once pest populations reach economic thresholds demand-
ing treatment, there is little recourse except use of pesticides or,
where applicable, insect pathogens.

The application of theory to practical problems was stressed by our
hosts. This emphasis came from scientists, Revolutionary Committees,
and production workers. Much of the research discussed and observed
was adaptive research; some of the basic ideas originated in other
parts of the world. However, it was apparent that fundamental biological information had been available for adapting these studies to production in their field situations, some of which were unique.

One of the most important plant-protection resources in China is the
immense human labor force. This was exemplified exceedingly well at
the Tahsia (Big Sand) Commune near Canton. Several of the techniques
used for plant protection of rice were highly labor intensive. These
included removing by hand-labor insect egg concentrations around light

traps (also spot treatment with insecticides), water and crop management, rearing and release of *Trichogramma* egg parasites, production
and application of *Bacillus thuringiensis*, and use of about 220,000
small ducks for leafhopper and planthopper control. About 685 people
were involved in handling the ducks, herding them from paddy to
paddy, and penning and feeding them at night. The value of *Trichogramma*
release on several crops is generally recognized in the United States,
but rearing costs, much of it hand labor, often prohibit purchase and
release in sufficient numbers to be effective. Techniques, such as
these examples, are valuable inputs into integrated control programs.
One of the fundamental principles of integrated control is the use of
pest population suppressive measures which are not disruptive to the
agricultural ecosystem, particularly to the biological control agents
in the environment.

Several Chinese speakers, particularly Revolutionary Committee representatives, indicated that, in their opinion, their political system
is well suited to development of plant protection and integrated control programs. This stemmed chiefly from the large labor resource and
the fact that products (e.g., insecticides) are not patented, there is
no profit incentive or competition for sales, and information is
quickly available for adoption nationwide. The main incentive is
production. It is left largely to the provinces and masses to try
out newly available techniques of plant protection. If it proves to
be an improvement over existing practices, it can be rapidly adopted.
Often techniques are adapted to local conditions and availability of
production supplies. Techniques for producing *Bacillus thuringiensis*
varied somewhat at the commune production level. Host eggs for rearing *Trichogramma* varied in different regions according to what host
eggs were conveniently and cheaply available.

The specific techniques useful in integrated control programs are
discussed in detail in the crop sections of this report. This section
will briefly discuss the general development of integrated control.

ECOSYSTEM RESEARCH

The foundation of integrated insect control is a detailed knowledge of
the crop ecosystem. The crop is not a unit in isolation but is part of
an ecological system with a multitude of interacting factors. This
aspect of integrated control development was not pursued in detail by
the study team, but clearly, in the major crops studied (e.g., rice),
much information has been accumulated over the years and is fundamental
to current advances in their technology. Such ecological information
has provided advances in biological control of scale pests (e.g., in
citrus crops), and in interplanting of certain crops to augment biological control activity (e.g., interplanting strips of wheat in
cotton fields to augment natural enemy populations). The strips of
wheat reduced aphid populations on the cotton plants and reduced the
spiny bollworm, *Earias insulana* Biosduval, problem on this crop. Pest
hibernation and breeding sites can be identified and destroyed with
this type of information.

CULTURAL CONTROLS AND AGRONOMIC PRACTICES

Entomologists in China are well aware that crop production practices
are related to pest populations and that modifications in prevailing
practices may create changes in the plant and environment. Change
may aggravate or ameliorate certain pest problems or even create
new ones, and studies are made to this point when changes in production practices occur. For example, many southern areas are now producing three crops of rice per year instead of two, and the impact
of this practice is being studied in relation to insect problems.
When two crops of rice were produced in Hunan Province, the rice stem
borer, *Chilo suppressalis* Walker, was the main pest, but time of planting and harvest under the triple-cropping system reduced the stem
borer populations, and the rice paddy borer, *Tryporyza incertulas*
Walker, became more important, as did the purplish stem borer, *Sesamia
inferens* Walker, in some localities.

Many modifications in crop production practices are designed to
give cultural control of certain pests. These often provide the first
line of defense against pest species, and other suppressive measures
(e.g., pesticides) are only supplementary. This is recognized
worldwide. In the labor-intensive agricultural system, cultural controls are particularly useful. Summer weeding eradicates transitional
hosts of leafhoppers and planthoppers affecting rice production and
reduces virus transmission. Good horticultural practices and winter
pruning enhance citrus tree vigor and reduce pest problems such as
those caused by longicorn borers. The Chinese have learned that flooding rice fields before March 6, just before emergence of rice paddy
borer from overwintering so reduces populations that insecticides are
not needed until the third generation and that sometimes no supplementary measures are necessary. It is possible that routine hand
pruning of vegetative branches and eventually "topping" cotton plants
may result in reduced plant attraction of *Heliothis* for oviposition
sites, but this feature has not been investigated to our knowledge.

Successful utilization of cultural controls requires a detailed
knowledge of the pest's biology and ecology in order to couple the
techniques with agronomic practices. Many successful cases of cultural control are given in the crop sections of this report. Certainly they are being used wisely and effectively with a concomitant
reduction in reliance upon insecticide use. The delegation was greatly impressed with the organization and success of this often-neglected
facet of integrated control.

ENVIRONMENTAL CONTROL OF LOCUSTS

The age-old locust problem in China remains severe but is being reduced
by sophisticated management techniques. Traditional scouting and forecasting of outbreaks is practiced to allow timely elimination with insecticides, but, significantly, an effort is being made to reduce
problems at the source through water management, reforestation, modi-
fying agricultural practices, and much more in a unified systems
approach.

The locust program provides an elegant example of successful insect control through major environmental changes. *Locusta migratoria* Linne is a pest in the eastern region of Shantung Province. Infestations date back to 1700 B.C. Recently there have been outbreaks every 3 to 5 years. The area around Weishan Lake, Shantung, can be divided into four zones: shallow water near the lake, flood plain, swampy area, and periphery. Each may be flooded or dry, depending on rainfall and water table. Locust infestations normally develop in the flood plain, and with high water table, in the swampy area and the periphery. Beyond the periphery, some areas may have infestations but only in years with extremely heavy rainfall. In the past, farmers suffered from alternating floods, droughts, and locust infestations.

Since 1950, hydrological engineering projects have been initiated, and monitoring and forecasting systems for locust populations have been instituted. In 1958, the physiography of the area was analyzed. In the 1960's, with 40 million man-days of labor and 550 million m³ of concrete, 423 irrigation systems with 21,552 ditches were built with a total length of 15,492 km. This hydrological system prevented flooding in 39,000 hectares (60,000 mu), even with 30 mm of rainfall, and a good harvest followed in 370,500 hectares (5,700,000 mu) despite 100 rainless days. In 1973, 58,890 hectares (910,000 mu) were planted to rice, 100,000 of which were formerly in a locust district. Also, 8,190 hectares (125,000 mu) around rice fields are planted to shade, fruit, and medicinal trees. These plantings reduced locust oviposition and increased niches for frogs and other natural enemies. After the water level is stabilized, 7,800 hectares (120,000 mu) will yield fish, and water vegetables.

Also in the 1960's, detailed studies of locust biology formed the basis of a program to eliminate the insects (Figure 15). The interactions of factors involved in controlling locusts through modification of the environment are given in Figure 2.

The program largely eliminated locust infestations in the region, reclaimed hundreds of hectares of arable land, and improved the general environment of the countryside. A monitoring and forecasting network is maintained to provide surveillance of possible local outbreaks.

A similar program in Hopei Province is also very successful. Fongnan County was an infested area. Since 1949, six reservoirs and several hundred irrigation ditches have been built that can drain water from 385 km². This construction reclaimed 11,700 hectares (180,000 mu) of land. Also, 3 million trees were planted. As a result, the area needing locust treatment is reduced to 2% of the pre-1949 area.

This summary is prepared on the basis of a presentation given at the Peking Institute of Zoology, Academia Sinica, Peking, and upon articles in Acta Entomologica Sinica:

1) Bureau of Agriculture and Forestry, Fongnan Hsien, Hopei Province, 1974. With line education as the key link, to change the status of the locust infested areas by means of combined control measures. Acta Entomologica Sinica 17:241-246.

2) Geening Bureau of Agriculture, Shantung Province & Laboratory in Insect Ecology, Peking Institute of Zoology, Academia Sinica. 1974.

146

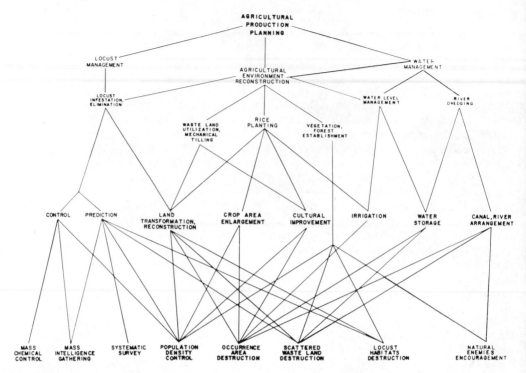

FIGURE 15 Pest management scheme for *Locusta migratoria* control in Weishan Lake District, Shantung Province. Figure translated and redrawn by Dr. Chang-Shyan Chang after Acta Entomologica Sinica *17(3)*, 247-257. 1974.

Elimination of locust infestation by combined efforts of reconstruction and chemical control in Weishan Lake district. Acta Entomologica Sinica *17*:247-257.

ECOSYSTEM DIVERSITY

Crop monocultures are sometimes damaged more severely by pests than the same crop grown in an area with crop diversity. Consequently, the assumption is often made that maximum diversity in an agricultural area is always desirable. It has been demonstrated that this assumption is not always valid; in many situations, diversity aggravates rather than ameliorates certain pest problems.

In most of the agricultural areas visited in China, the fields were relatively small, and several crops were produced in each area. Rice was a major exception, particularly in the southern provinces; vast areas were uniformly planted to this crop. In southern areas, rice is often triplecropped annually, but farther north it commonly is grown in rotation, e.g., corn-rice-vegetables or rice rotated with one other crop.

As nearly as could be determined, not a great deal of research attention has been given to intensification of pest problems or to amelioration through agroecosystem diversity. In northern areas, for example, sorghum, millet, and corn are all grown in neighboring fields, and it was noted that pest problems associated with these crops are similar. Quite likely, the need to produce food crops in maximum quantities reduces the practicality of diversifying the cropping pattern. A type of diversity was noted, however, where strips of wheat were planted through cotton fields. Populations of lady beetles build up on the wheat and move to the cotton as the wheat matures. This occurs at a time when aphid populations may become a problem on the cotton.

PEST AND BENEFICIAL ARTHROPOD COMPLEX

A large complex of insect pests are on the crops grown in China. Many crops are also seriously affected by spider mite species, particularly deciduous fruit and citrus. The insect pest situation is generally similar to that in important agricultural areas worldwide in that the pest complex on each crop include a few species of consistently serious pests, a larger group of secondary pests of occasional importance, and many pests that are seldom serious but could become so.

The delegation observed and discussed many crops and associated pest problems with their hosts, but attention was concentrated on major crops, principally food crops and cotton. The details of these plant protection problems and solutions are discussed in Chapter 3, but it is obvious that a detailed knowledge of insect pests and their interactions within the crop and environment is available in China.

The situation regarding beneficial insects was similar. Important parasites and predators have been identified, and the relative significance of each beneficial species in the various crop ecosystems is well understood if not quantified in many cases. Quantification for more than a few intensively studied species is rare in any country. Entomologists and apparently many workers at the production brigade level are aware of the value of beneficial species and are seeking means to conserve and augment their populations.

Advances in this activity have been steady, although some areas visited seemed to be more dedicated and advanced than others in work with biological control organisms.

Efforts directed toward augmentation have been outstanding in the case of mass propagation and release of egg parasites, particularly several species of *Trichogramma* and, to a lesser extent, *Anastatus* sp. Release rates are sufficiently high to be consistently effective, a factor that has proved prohibitively expensive in the United States in many cases. Further, careful studies of the biological parameters of each species of *Trichogramma* have identified the environments and pest species for which each is most suitable for mass release in the various regions and crops of China.

Other rather sophisticated management of biological control species is described in the report. There have been shipments of biological control species between the provinces, but the response to questions by the delegation regarding shipment of exotic species into China did not encourage pursuance of this matter further. The political theme of "self reliance"--interpreted as "Do it with what we have or without outside assistance"--apparently discourages international collaboration of this nature at this time. It was the opinion of the delegation that plant protection in China could profit by importation of carefully selected beneficial insects. The delegation is aware of the failures and frustrations that often result, but in some cases this could yield immediate tangible gains, such as importation of predaceous mites resistant to organophosphate insecticides, possibly species such as *Metasieulus occidentalis* or *Amblyseius fallacis,* along with the management technology to preserve them in deciduous fruit orchards. Also, China probably has beneficial species of importance to crop protection in the United States, e.g., parasites of the red scale of citrus in the two interior areas of California.

USE OF PESTICIDES IN INTEGRATED CONTROL PROGRAMS

On several occasions, the Chinese told the delegation how completely they had come to rely on insecticides. Although data was not furnished on amounts applied, they expressed deep concern regarding compounding problems resulting from this heavy dependence. It is widely accepted that insecticides are the main contributor to destruction of beneficial insect populations. The resulting loss of biotic regulations causes familiar problems of spiraling production costs, intensification of insecticide resistance, and increasingly severe pest problems.

Although a continuing need for insecticides is recognized, reevaluation of plant protection programs led the Chinese to the integrated control approach. To this end, they have concentrated on use of insecticides with a moderate to high degree of selectivity. Use of the organochlorine chemicals is being phased out (as well as use of the most hazardous insecticides). They have concentrated on use of physiologically selective chemicals and on using them in an ecologically selective manner. Efforts were also intensified toward development of tactics alternative to the use of chemicals (cultural controls, etc.). The success of these programs varied somewhat in areas visited. Obviously, insecticides are still widely used, but use now is something of a last resort based on forecasting and use of economic thresholds; when pest populations surpass the economic threshold, insecticides are applied.

The success of applying integrated control practices was strikingly illustrated at the Tahsia (Big Sand) Commune in Kwangtung Province. The development of techniques alternative to the use of synthetic organic insecticides and careful pest forecasting has resulted in a decrease in the insecticide load on rice from 75,427 kg in 1972 to 34,112 kg in 1974 (a reduction of more than 50%), and it appeared that reductions in use would be considerably greater in 1975 (only 6,714 kg had been used at the time of our visit in late August). Other areas visited obviously had reduced their use of insecticides significantly, but data were not available as to quantity.

USE OF ENTOMOPATHOGENS IN INTEGRATED CONTROL PROGRAMS

Details of research and utility in field programs are given in the report section on biological control. *Bacillus thuringiensis* is used widely throughout China, *Beauveria* has been developed for use only in recent years and is used less widely primarily for corn borers, and baculoviruses are beginning to be used for several species of foliage-feeding lepidopterous larvae. The use of entomopathogens has been popularized since they have proved efficient, are not costly in their system, and present little hazard to workers; moreover, they can be produced at the brigade level in the communes.

Entomologists in China are enthusiastic regarding their successful use of entomopathogens. These organisms epitomize the selectivity so difficult to attain with the synthetic insecticides. They are believed to be almost totally nontoxic to beneficial species. As a consequence, their use in regulating pest populations is an important tactic in integrated control systems.

Use of these insect diseases is often combined with other measures. Sometimes this will be with another organism such as *Trichogramma* or, if pest outbreaks are unusally severe, they may be combined with a minimal dosage rate of an insecticide.

The delegation was greatly impressed with the use of insect diseases as a control tactic, and our limited field observations confirmed the good results described by our hosts. The use of *B.t.* in the field generally, and of *Beauveria* in Kirin Province may be more intensive than in any other country in the world. Consequently, selection pressures for the development of resistance of these pathogens are more intense than elsewhere. Entomologists have long theorized about the ability of insects to develop resistance to insect diseases. It will be interesting and certainly informative if resistant strains of insects are eventually selected in China.

PEST POPULATION FORECASTING

The well-coordinated and effective programs of pest forecasting are described under "Extension Programs" in Chapter 1. The programs coordinate activities at the province, district, county, and commune levels. Information on pest abundance, population trends, and control measures is disseminated. Information on many of the crops studied is sufficient to enable entomologists to establish economic thresholds (levels of pest abundance at which control measures are indicated). There was no way to ascertain the accuracy of these thresholds; in most parts of the world, thresholds are continually refined as further information becomes available. From our information, however, they seem to be serving well, preventing undertreatment, with economic losses resulting, and overtreatment, with concomitant unnecessary costs and biological problems.

STERILIZATION AND THE STERILE MALE RELEASE TECHNIQUE

In our trip through China, no mention was made of the development of the sterile-male release method for any insect pest. One of our hosts indicated that in the 1960's a screening program was undertaken at Peking University to find effective chemosterilants. An institute in Canton that we did not visit was mentioned as being interested in research on chemosterilants. Originally, the goal of chemosterilant screening was related to armyworm control. Apparently current interest is more closely related to control of the rice paddy borer.

Research at the Department of Plant Protection, Kwangtung College of Agriculture and Forestry, was mentioned, but that institution was not on our agenda and was not visited. Their work on the effects of irradiation and chemosterilization of the rice paddy borer, *Tryporyza incertulas* Walker, was published in 1974[1]. Laboratory and field tests showed that old pupae exposed to 35,000 r from a ^{60}Co source produced adults that were completely sterile and had normal longevity and mating capacity. Tests with thiotepa were also promising. Male moths are completely sterilized by exposure of 1 minute to the residue resulting from a mixture containing 16% thiotepa (20.9 µg per milimeter^2), without adverse effect on longevity and mating frequency. In field tests, 25 male moths which had contact with the chemosterilant and 25 untreated females were released in a large screen cage in a rice seedling field. In another cage, 25 pairs of untreated moths were released. The moths were free to mate and oviposit on the rice seedlings. After 17 days, the percentage of dead-hearts and the number of infertile eggs in each egg mass were recorded. There were 1.3% dead-hearts in the treated cage and 85% in the control cage. Twenty-eight percent of the eggs were infertile in the treated cage, but only 1.7% were infertile in the control cage. Further field tests were proposed.

STATUS OF SOME OTHER TECHNIQUES: HOST PLANT RESISTANCE, PHEROMONES

The status of the principal techniques used in China has briefly been discussed. Various light and bait traps are being used; light trapping is widespread. Research on pheromone trapping is progressing rapidly, but more must be done before it is used widely as a population prediction tool, although preliminary studies are under way to study population buildup, migration, and so on. There is great interest in the potential of pheromones as a population control through interruption of mating communication.

There is interest in breeding for host-plant resistance to insect attack. Most research attention has been given to improvement of yield and quality of crops but includes breeding for resistance to plant pathogens. At present, however, attention to programs for breeding for resistance to insect attack is minimal, certainly less than desirable. Teams of plant breeders, entomologists, and other research people have not been formed to our knowledge.

The delegation agreed that interest in and support for plant protection and particularly in development of integrated control was very impressive in China. It has a much higher priority than in the United States and perhaps more than in any other part of the world. It is apparent that the high priority placed on agricultural production and plant protection has brought steady, sometimes spectacular, progress on all agricultural fronts, both in production and in protection.

REFERENCE

1. The Teaching and Research Group of Entomology and Pesticides, Department of Plant Protection, Kwangtung College of Agriculture and Forestry. 1974. A preliminary study on the application of the sterility technique for the eradication of the paddy borer, *Tryporyza incertulas*. Acta Entomologica Sinica *17*:135-147.

Appendix A

ITINERARY OF AMERICAN

INSECT CONTROL DELEGATION

PEKING

Monday, August 4

5:05 p.m.	Arrival in Peking
6:00 p.m.	Peking Hotel
7:30 p.m.	Itinerary discussions: Guyer, Metcalf, Beemer

Tuesday, August 5

8:30 a.m.	Peking Institute of Zoology, Academia Sinica
	Lectures:
	Meng Hsiang-ling
	Kung K'un-yüan
	Chou Hou-an
	Ma Shih-chün
	Ch'en Ning-shen
5:00 p.m.	Reception by Ambassador Bush at U.S. Liaison Office

Wednesday, August 6

9:00 a.m.	Peking Institute of Zoology, Academia Sinica
	Lectures:
	Ch'en Te-ming
	Liu Meng-yin
	Ts'ai Hsiu-yü
2:00 p.m.	Peking Institute of Zoology, Academia Sinica Visit to laboratories
7:00 p.m.	Welcoming banquet by STA hosted by Chou Pei-yuan, Vice-Chairman of the Scientific and Technical Association of the People's Republic of China

CHANGCHUN, KIRIN

Thursday, August 7
9:00 a.m.   Leave Peking by air for Changchun
12:40 p.m.   Arrive Changchun
2:00 p.m.   Kirin Institute of Applied Chemistry,
                    Academia Sinica
                    Lecture:
                    Shen Lien-fang
7:00 p.m.   Welcome and banquet hosted by Chang Shih-ying,
                    Member of the Standing Committee of the
                    Revolutionary Committee of Kirin

KUNGCHULING, KIRIN

Friday, August 8

7:30 a.m.   Leave Changchun for Kungchuling
9:00 a.m.   Arrive Kirin Academy of Agricultural Sciences
                    Lectures:
                    Hsü Ch'ing feng
                    Ch'en Jui-lu
                    Hsü En-p'ei
11:30 a.m.   Discussion with plant breeders
                    Participants: Adkisson, Chiang, Maxwell
2:00 p.m.   Nan Weitzu Commune
                    Scientific and Technical Station of
                    Ta Yushu Production Brigade
3:30 p.m.   Nan Weitzu Commune
                    Beauveria Production Brigade
8:00 p.m.   Delegation members' lectures at the Kirin
                    Institute of Applied Chemistry:
                    Adkisson, Roelofs

CHANGCHUN, KIRIN

Saturday, August 9

8:30 a.m.   Kirin Institute of Forestry Research
                    Lectures:
                    Li Wang-hung
                    Yü En-yu
2:00 p.m.   Sightseeing in Changchun
4:15 p.m.   Flight to Peking
8:05 p.m.   Arrive Peking

PEKING

Sunday, August 10

9:30 a.m.	Sightseeing at the Great Wall
12:30 p.m.	Sightseeing at Ming Tombs
4:00 p.m.	Shopping at Friendship Store
7:30 p.m.	Delegation group meeting

Monday, August 11

9:00 a.m.　　Peking Institute of Zoology, Academia Sinica
　　　　　　　　Lectures:
　　　　　　　　Guyer
　　　　　　　　Chiang
　　　　　　　　Huffaker
　　　　　　　Rest of delegation to Peking Zoo
2:30 p.m.　　Szechiching (Evergreen) People's Vegetable
　　　　　　　　Commune
8:00 p.m.　　Delegation group meeting

Tuesday, August 12

8:30 a.m.　　Peking University
　　　　　　　　Entomological Laboratory visit
　　　　　　　　Lecture:
　　　　　　　　Lin Ch'ang-shan
2:30 p.m.　　Peking Institute of Zoology, Academia Sinica
　　　　　　　　Lectures:
　　　　　　　　Metcalf
　　　　　　　　Reynolds
　　　　　　　　Weidhaas
　　　　　　　Rest of delegation to Palace Museum
7:30 p.m.　　Lectures at Peking Hotel
　　　　　　　　Lectures:
　　　　　　　　Maxwell
　　　　　　　　Roelofs
　　　　　　　Discussions with Chinese Agricultural
　　　　　　　　Association at Peking Hotel
　　　　　　　　Participants: Guyer, Metcalf

Wednesday, August 13

8:30 a.m.	Peking Vegetable Market
10:30 a.m.	Summer Palace
2:00 p.m.	Peking Zoo (Guyer, Huffaker)
	Palace Museum (Metcalf, Reynolds, Weidhaas)
	Free time (rest of delegation)

Wednesday, August 13 (continued)

6:30 p.m.   Banquet for Chinese hosts
                  Chengtu Restaurant

SIAN, SHENSI

Thursday, August 14

9:50 a.m.   Leave Peking for Sian
2:25 p.m.   Arrival in Sian
3:00 p.m.   Sightseeing in Hwa Ch'ing Shih
8:00 p.m.   Puppet show

WUKUNG, SHENSI

Friday, August 15

9:00 a.m.   Arrive Northwest College of Agriculture
                     Lectures:
                     Yüan feng
                     Ma Ku-fang
                     Sun Yi-chih

SHANGHAI

Saturday, August 16

9:00 a.m.   Leave Sian by air for Shanghai
2:00 p.m.   Arrive Shanghai
7:30 p.m.   Delegation itinerary meeting

Sunday, August 17

9:a.m.   Shanghai Industrial Exhibition
2:30 p.m.   Sightseeing: Shanghai Building, Friendship
                     Store, Waterfront

Monday, August 18

8:45 a.m.   Tour of Tsaoyang Workers' New Village
2:30 p.m.   Shanghai Academy of Agricultural Sciences
8:00 p.m.   Films at hotel

Tuesday, August 19

9:00 a.m.	Shanghai Institute of Entomology
2:00 p.m.	Shanghai Institute of Entomology
7:00 p.m.	Banquet hosted by Hsü Yen, Responsible Member of the Shanghai STAPRC

Wednesday, August 20

8:45 a.m.	Shanghai Children's Palace
2:30 p.m.	Changchen (Long March) Vegetable Commune
7:15 p.m.	Acrobatic performance

Thursday, August 21

8:30 a.m.	Agricultural Information Exchange Center
	  Lectures:
	  Chiang
	  Reynolds
	  Roelofs
	Nanshang Village Patriotic Hygiene Unit
	  (rest of delegation)

2:30 p.m.	Shanghai Institute of Organic Chemistry, Academia Sinica (Metcalf, Roelofs, Beemer) Shanghai Agricultural Exhibition (Guyer) Tangwan Commune (rest of delegation)
8:30 p.m.	Meeting with representatives of Shanghai Agricultural Pesticide Factory

CHANGSHA, HUNAN

Friday, August 22

7:50 a.m.	Leave Shanghai for Changsha
2:00 p.m.	Sightseeing in Changsha
7:30 p.m.	Banquet hosted by Sun Yun-ying, Member of the Standing Committee of Hunan Provincial Revolutionary Committee

Saturday, August 23

8:30 a.m.	Hunan Institute of Plant Protection

2:30 p.m.	Hunan Institute of Plant Protection
	  Lectures at hotel:
	  Lei Hui-chih
	  Ch'en Ch'ang-ming

SHAO-SHAN, HUNAN

Sunday, August 24

8:30 a.m.	Sightseeing in Shao-shan
6:50 p.m.	Leave by train for Kwangchow

KWANGCHOW, KWANGTUNG

Monday, August 25

7:30 a.m.   Arrive Kwangchow
10:30 a.m.   Foshan City Patriotic Hygiene Movement Unit
2:00 p.m.   Hsinhua Production Brigade, Shun-te County
           Lectures:
           Huang Tzu-jan
           Yang Tsan-hsi

Tuesday, August 26

8:00 a.m.   Chungshan University
           Lecture:
           P'u Che-lung
2:30 p.m.   Kwangtung Institute of Entomology
           Lectures:
           Mai Hsiu-hui
           Wang Ting-hsiang
           Li Li-ying

Wednesday, August 27

9:30 a.m.   Tahsia (Big Sand) Commune, Sse-hui County
           Lectures:
           Ku Te-hsiang
           Luo Yu-ch'uan
2:00 p.m.   Visit Tahsia: Commune cropping areas
           Microbial Control Agent factory

Thursday, August 28

8:30 a.m.   Sightseeing in Kwangchow
9:30 a.m.   Visit to Kwangtung Botanical Garden,
           Institute of Botany (Guyer, Metcalf)
2:30 p.m.   Lectures at Science and Technology Hall:
           Guyer, Metcalf, Huffaker, Weidhass, Roelofs

6:00 p.m.   Banquet
8:30 p.m.   Party for STAPRC accompanying staff

Friday, August 29

9:30 a.m.   Leave Kwangchow by train for Hong Kong
3:30 p.m.   Arrive Hong Kong

Appendix B

UNITS VISITED AND PEOPLE MET

The Scientific and Technical Association of the People's Republic of China

As the hosting organization, the Scientific and Technical Association of the People's Republic of China (STAPRC) was instrumental in the success of the delegation's trip. While primary responsibility for this success rests with the group that accompanied the delegation throughout the visit, at each stop, the local STAPRC office provided invaluable itinerary and logistical support. Besides the four persons who travelled with the delegation (named in the Preface) the delegation wishes to thank the following persons for their assistance:

PEKING

周培源   Chou P'ei-yüan
    Vice-Chairman
    National Science Association

朱永行   Chu Yung-hang
    Vice-Director
    Foreign Affairs Bureau, National Science Foundation

冯因复   Feng Yin-fu
    Deputy Chief of Division
    Foreign Affairs Bureau, National Science Association

CHANGCHUN, KIRIN

吴云超   Wu Yun-shao
    Staff Member
    Kirin Provincial Scientific and Technical Association

160

SHANGHAI

许 言　　Hsü Yen
　　　　　Responsible Member
　　　　　Shanghai Scientific and Technical Association

毛生芳　Mao Shen-fang
　　　　　Office Responsible Member
　　　　　Shanghai Scientific and Technical Association

陆关虎　Lu Kuan-hu
　　　　　Officer
　　　　　Shanghai Scientific and Technical Association

蔺新耕　Shen Chin-chung
　　　　　Officer
　　　　　Shanghai Scientific and Technical Association

沈金荣　Lan Hsin-keng
　　　　　Officer
　　　　　Shanghai Scientific and Technical Association

夏安之　Hsia An-chih
　　　　　Responsible Member
　　　　　Shanghai Scientific and Technical Association

许鼎祥　Hsü Ting-hsiang
　　　　　Worker
　　　　　Shanghai Scientific and Technical Association

丁益山　Ting Yi-shan
　　　　　Worker
　　　　　Shanghai Scientific and Technical Association

顾永昌　Ku Yung-ch'ang
　　　　　Worker
　　　　　Shanghai Scientific and Technical Association

CHANGSHA, HUNAN

田后凤　T'ien Hou-feng
　　　　　Vice-Director
　　　　　Hunan Province Science and Technology Bureau

丁昌浩　Ting Ch'ang-hao
　　　　　Office Chairman
　　　　　Hunan Province Science and Technology Bureau

王德广　Wang Te-kuang
　　　　　Worker
　　　　　Hunan Province Science and Technology Bureau

KWANGCHOW, KWANGTUNG

 Kuo Ch'ang
Member
Shaoshe Area Revolutionary Committee
Scientific and Technical Bureau

The Peking Institute of Zoology, Peking

The Peking Institute of Zoology, which is under the dual jurisdiction of the Academia Sinica and the Peking Municipal Revolutionary Committee, was founded in 1962 when the Institute of Entomology joined the Institute of Zoology. It is divided into 10 departments: 5 in entomology and 5 in zoology. There are about 500 research workers at the Institute. Between 40 and 50 of these are "experienced workers" (the equivalent of associate or full professor), about 250 are "younger experimental science members," and about 200 are staff members and factory workers. The institute had two factories attached to it as part of its operations, one formerly producing insecticides (including microbiological insecticides) and the other producing animal hormones, but now the factories have been combined and are used for repairing scientific instruments used at the Institute.

The 1:5 ratio of experienced research workers to younger experimental workers is highly unusual; more often, institutes have ratios of 1:100-200. The institute is an old one with a long history of research work and has, therefore, a disproportionately large number of experienced research workers.

The research workers from this institute attend two special May 7 Cadre Schools that are run by the Academia Sinica on the outskirts of Peking. The personnel from the institute attend the sessions at the Cadre Schools for periods of 6 months to a year, returning to Peking once a month to spend weekends at home.

The organization of the Peking Institute of Zoology is as follows:

ADMINISTRATION
Administrative Office, Business Office, Political Office

RESEARCH UNITS
Entomology: Insect Taxonomy, Insect Ecology, Physiology, Insect Pathology, Insecticides and Toxicology
Zoology: Invertebrate Taxonomy, Vertebrate Taxonomy, Cytology, Hormones, Techniques

陈世骧 Ch'en Shih-hsiang
Director

朱弘复　Chu Hung-fu
　　　　　Vice-Director

马世骏　Ma Shih-chün
　　　　　Laboratory Director
　　　　　Insect Ecology Laboratory

钦俊德　Ch'in Chün-te
　　　　　Laboratory Director
　　　　　Insect Physiology Laboratory

龚坤元　Kung K'un-yüan
　　　　　Responsible Member, Business Office
　　　　　Laboratory Director, Insecticidal Toxicology
　　　　　Laboratory

熊　尧　Hsiüng Yao
　　　　　Laboratory Vice-Director
　　　　　Insecticidal Toxicology Laboratory

陈德明　Ch'en Te-ming
　　　　　Laboratory Director
　　　　　Technical Laboratory
　　　　　Head, Insect Pheromone Section

刘孟英　Liu Meng-ying
　　　　　Researcher

孟祥玲　Meng Hsiang-ling
　　　　　Researcher

周厚安　Chou Hou-an
　　　　　Researcher

陈宁生　Ch'en Ning-sheng
　　　　　Researcher

蔡秀玉　Ts'ai Hsiu-yü
　　　　　Researcher

Kirin Institute of Applied Chemistry, Changchun, Kirin Province

This institute in Changchun is under the dual direction of the Academica Sinica and the Revolutionary Committee of Kirin Province. It was established in 1948 as Kirin Polytechnic Institute, and in 1953 the Chemistry Department became affiliated with the Institute of Physical Chemistry of Shanghai to form the present organization. This is a large institute with about 1,100 research workers and assistants.

马祖沿　Ma Tsu-chih
　　　　　　Vice-Chairman
　　　　　　Revolutionary Committee

沈联芳　Shen Lien-fang
　　　　　　Responsible Member
　　　　　　Structural Analysis Research Laboratory

赖光赐　Lai Kuang-szu
　　　　　　Secretary
　　　　　　Science and Technology Office

张维纲　Chang Wei-kang
　　　　　　Researcher
　　　　　　High Polymer Physics
　　　　　　Interpreter

朱贵发　Chu-Kuei-fa
　　　　　　Researcher
　　　　　　Inorganic Analysis
　　　　　　Interpreter

Kirin Academy of Agricultural Sciences, Kungchuling, Kirin Province

Located at Kungchuling, Kirin Province, the Kirin Academy of Agricultural Sciences (KAAS) was founded in 1949 under the name of the Northeast Agricultural Research Institute. In 1959, it was renamed the Kirin Academy of Agricultural Sciences. Its purpose is to help increase crop and animal production in the province.

There are 779 employees working for KAAS, of whom 256 are research workers who are scientists or have science training. On a 3-year rotating basis, one-third of these scientists work full time at KAAS headquarters, one-third are located at headquarters but are on call to provide technical assistance to the 296 local field experimental groups, and one-third are located at 29 commune- and brigade-level agricultural research stations where they cooperate with local technicians. In this way, the KAAS is an integral part of a province-wide research, extension, and production team, and all scientists at KAAS headquarters obtain practical experience on field problems. Similarly, the KAAS is made aware of field problems and has a ready mechanism for disseminating research results and other information to farmers.

There are 400 hectares of cultivated land, some 100 of which are in field test plots. Five research institutes and an experimental farm are under the administration of KAAS: Crop Breeding Research Institute, Institute of Soils and Fertilizer (and Crop Management), Institute of Plant Protection, Institute for Animal Husbandry, Institute for Fruit Improvement, and the experimental farm.

The KAAS participates in a research-extension-production system involving the counties, the communes, the production brigades, and the production teams. KAAS scientists (85 of them) are stationed at 29 agricultural research stations set up at the commune and production levels. They assist technicians in carrying out field testing of materials from KAAS as well as selections found locally. Another 85 KAAS personnel, while stationed at Kungchuling, assist technicians at 296 field testing sites focused primarily on the production team level. This system provides the linkage with production.

苏群　Su Ch'ün
　　　　Vice-Chairman

李义忠　Li Yi-chung
　　　　Vice-Chairman

胡吉成　Hu Chi-ch'eng
　　　　Chairman
　　　　Institute of Plant Protection

何庸　He Yüng
　　　Vice-Chairman
　　　Institute of Plant Protection

徐恩培　Hsü En-p'ei
　　　　Section Leader, Insect Station
　　　　Institute of Plant Protection

徐庆丰　Hsü Ch'ing-feng
　　　　Researcher
　　　　Biological Control of Corn Borer with Beauveria

王承倫　Wang Ch'eng-lun
　　　　Researcher
　　　　Biological Control of Corn Borer with Trichogramma

玉芸生　Wang Yün-sheng
　　　　　Researcher
　　　　　Biological Control of Corn Borer with Sex Attractants

陈瑞鹿　Ch'en Chui-lu
　　　　　Head
　　　　　Department of Pest Forecasting

桂承明　Kui Ch'eng-ming
　　　　　Researcher
　　　　　Control of Soil Insect Pests

　　　　　Chang Tse-ching
　　　　　Researcher
　　　　　Soybean Breeding

　　　　　Hsieh Tao-hung
　　　　　Researcher
　　　　　Corn Breeding

　　　　　Wang Chin-hsien
　　　　　Researcher
　　　　　Wheat Breeding

　　　　　Wang Fang
　　　　　Researcher
　　　　　Sorghum Breeding

　　　　　Wang She-i
　　　　　Researcher
　　　　　Sorghum Breeding

时俊峰　Shih Chün-feng
　　　　　Vice-Chairman
　　　　　Huai-te County Revolutionary Committee

刘敏政　Liu Min-cheng
　　　　　Office Vice-Chairman
　　　　　Huai-te County Revolutionary Committee

Ta Yushu Production Brigade Science and Technical Station of Nan Weitzu Commune, Kungchuling, Kirin Province

This brigade is one of 11 production brigades that make up the Nan Weitzu Commune. The brigade has a population of 3,700 people and farms 480 hectares of land, about 100 hectares of which are in paddy rice. Corn, sorghum, and soybeans are raised. The total grain production for 1974 was 3,100 metric tons for an average of 8.122 kilos per hectare.

The plant production work done in the brigade is closely linked to the research done at KAAS. The KAAS laboratories produced the first strains of *Beauveria* in the fight against the European corn borer. Currently, the *Beauveria* is cultivated at the commune-level Workshop for Beauveria Production.

刘国臣　　Liu Kuo-ch'en
　　　　　　Chairman
　　　　　　Revolutionary Committee

牛国戎　　Niu Kuo-chung
　　　　　　Vice-Chairman
　　　　　　Revolutionary Committee

孙中臣　　Sun Chung-ch'en
　　　　　　Technician,
　　　　　　Agricultural Experiment Station

Workshop for Beauveria Production of Nan Weitzu Commune, Kungchuling, Kirin Province

This workshop is associated with the KAAS and is the basic production unit for the cultivation of *Beauveria* in Nan Weitzu Commune. Seven staff members work at the establishment in December and January and then March through July. The establishment has a production capacity of 5 kilos of second-generation *Beauveria* a day. When diluted at a 10:1 ratio, this amount produces 50 kilos a day for field use. The yearly production of this diluted third-generation *Beauveria* is 1.5 metric tons.

Kirin Institute of Forestry Research, Changchun, Kirin Province

Under the leadership of the Kirin Branch of the Ministry of Agriculture and Forestry, this institute has been involved in research and applied work since its inception in 1954. Its basic tasks are solving forest production problems, popularizing knowledge in forestry, training technicians who return to production units, and conducting some basic research. It is divided into the departments of forestation, forest management, forest protection, and forest multiple use and production. This unit does research on both provincial and national levels.

One of the main problems the institute has been involved in recently is the pine caterpillar, which infests the pine, larch, and related forests of the province. Since 1964, it has been involved in a program of pine caterpillar control using *Trichogramma*. Through the use of

insect control stations throughout the province, the program has virtually eliminated the pest problem in Kirin Province.

周云峰   Chou Yün-feng
            Responsible Member,
            Revolutionary Committee

何平勛   Ho P'ing-hsün
            Responsible Member,
            Protection Laboratory

余恩裕   Yü En-yü
            Responsible Member,
            Insect Division

李旺鱟   Li Wang-hung
            Technician,
            Protection Laboratory

衣振山   Yi Chen-shan
            Technician,
            Protection Laboratory

时荣元   Shih Chung-yüan
            Technician,
            Protection Laboratory

Szechiching (Evergreen) Commune, Peking

The Szechiching (Evergreen) Commune is about 40 minutes by car from
central Peking and is northwest of the city, near the Summer Palace and
Peking University. The commune was founded in August 1958 from six high-
level agricultural cooperatives. It has 14,000 households or 42,000
people divided into 14 production brigades and 143 production teams, and
occupies 2,734 hectares of land. The commune also has six factories and
a vegetable-seed experiment station. Water conservation work has pro-
ceeded since the commune's founding, and Evergreen is now 98% irrigated,
with more than 580 electrically pumped wells. The commune owns 800
pieces of machinery that includes more than 140 trucks, 70 tractors,
300 hand tractors, some combines, and well-sinking and other machines.
The commune primarily produces vegetables but also grain, fruit, pigs,

and ducks. An area of 1,467 hectares is in vegetable production, and
the yields have risen annually. The commune is now supplying the state
120 million kg of vegetables each year, of over 100 varieties (up from
50 before 1966) and in all seasons. Commune land of 867 hectares is
used for grain production, mainly wheat, rice, corn, and sorghum. Annual yields have risen from 2,000 kg per hectare before 1958 to 6,000
in 1975. Four hundred hectares of Szechiching land are used to raise
fruit--primarily grapes, pears, apples, and peaches. The present annual yield of 6 million kg of fruit is more than 120 times the pre-1958
yield of the area. The commune also supplies the state with 17,000 pigs
and 27,000 Peking ducks annually.

Educational facilities consist of 18 primary schools and 7 middle
schools in the commune. In 1974 more than 40 commune residents went on
to university. There will soon be 35 students from Tsinghua University's
Machinery and Electric Parts Division at the commune, and a plant protection department of an agricultural college will also be established
at the commune. Progress in health included establishment of one hospital and 14 cooperative medical stations--one for each brigade. These
facilities have helped train 80 doctors and 200 "barefoot doctors."
The commune also has mobile movie-showing teams, wired radio, and a
telephone network. Farmers have private plots, limited to 1/20 mu per
person. The commune's 20,000 workers' average income is about $300
annually.

王 順 英　　Wang Shun-yin
　　　　　　　Vice-Chairman
　　　　　　　Commune Revolutionary Committee

李 文 华　　Li Wen-hua
　　　　　　　Cadre,
　　　　　　　Commune Revolutionary Committee

姜 淑 琴　　Hsi Shu-ch'in
　　　　　　　Responsible Member,
　　　　　　　Commune Revolutionary Committee

岳 春 玲　　Yüeh Ch'un-ling
　　　　　　　Cadre,
　　　　　　　Commune Revolutionary Committee

崔 凤 玉　　Ts'ui Feng-yü
　　　　　　　Technician,
　　　　　　　Commune Science and Technology Committee

朱文成　Chu Wen-ch'eng
Technician,
Commune Science and Technology Committee

Peking University, Peking

Peking University, founded in 1898, has 20 departments in the natural sciences, humanities, and technology. It has more than 5,000 students, and the number is expected to rise to 8,000 with the next class and to be over 10,000, the ultimate size, within a few years. There are 200 foreign students from some 20 countries. The university has very strong ties with 65 factories where education, research, and production are conducted on an extension basis. Professors are expected to spend about one-third of their time in such extramural activities.

周 培源　Chou P'ei-yüan
Vice-Chairman
Revolutionary Committee

陈 阅 增　Ch'en Yüeh-tseng
Vice-Chairman,
Biology Department
Revolutionary Committee

李 正 理　Li Chen-li
Professor,
Biology Department, specialist in plant morphology
and anatomy

林 昌 善　Lin Ch'ang-shan
Professor,
Biology Department

宗 志 祥　Tsung Chih-hsiang
Lecturer,
Biology Department

李 应 久　Li Ying-chiu
Office Cadre,
Revolutionary Committee
Responsible Member,
Revolutionary Committee

吴鹤龄   Wu He-ling
          Lecturer,
          Head of Cancer Research Team

蔡晓明   Ts'ai Hsiao-ming
          Biology Department

杨佟美   Yang Ling-mei
          Biology Department

杜之兰   Tu Chih-lan
          Biology Department,
          Entomological specialty

District Market, Peking

The market is large, consisting of two warehouse-type buildings, one
behind the other. Many vegetables and fruit stands were at the entrance
to the store. These stands remain open 24 hours a day, allowing
convenience-shopping when the main market is closed.

The market was exceptionally clean and orderly. The main items observed were vegetables, fruits, meats, cookies, candy, and drinks. In
general, shoppers brought shopping bags or baskets, so purchases needed
no wrapping. Meats and other products were wrapped in newspapers.
Costs of purchases were calculated by the abacus.

The principal meats available were fish, pork, lamb, chicken, and duck;
but there were hearts, liver, and sausage-type meats also. At the meat
counters, the meats were picked up by hand, but money taken by the clerk
and the change returned were picked up by forceps. Meats other than
duck, chicken, or fish were hung as carcasses behind the counter and
cut at the counter. Ducks were already cleaned and hung by the feet for
display. At one counter, small, live chickens in cages were selected,
their feet tied, and then deposited live into shopping bags. At another
counter, one could get the chicken killed and plucked.

This market is one of several similar district markets in Peking.
Most of its produce comes from a single production brigade (with eight
production teams) located in suburban Peking. Prices paid for the vegetables vary from season to season. Tomatoes that sold for 8 cents/
kilogram (kg) at the time of our visit in August, were 3.5 cents/kg in
July. They will rise to 60 cents/kg during winter. Each day, 10,000
customers buy $10,000 worth of produce, either at the main store between
7 a.m. and 7 p.m. or at the smaller after-hours store in front. Besides
vegetables, they buy 4.55 tonnes of pork (90-95 cents/kg), 1.8 tonnes

of fish (45 cents/kg), 600-700 chickens (45-50 cents/kg of live chickens), and 200 ducks each day. The market's meat comes from Peking Food Company, a slaughterhouse and packer that sells in response to daily requests, and its shrimp ($1.65/kg) comes from Peking Sea Product Company, which brings it frozen from the coast. Pricing is fixed according to standards set down by a Peking municipal government organization. Prices for some items drop if unsold by evening. (See Table B-1.) Meat could be held 1-2 months in a small refrigeration unit. Overall supply and demand are kept in line by shipping summer excess produce to the Northeast or Northwest.

Sanitation in the market was excellent. All areas were clean and free of flies. The manager indicated that the market was sprayed with dichlorvos each morning before opening.

Northwest College of Agriculture, Wukung, Shensi Province

The Northwest College of Agriculture was founded in 1934 and has just completed four decades of growth. Most of the expansion has occurred since 1949. The area has increased from 48 to 123 hectares, and the number of laboratories for research and teaching has increased from 3 to 84. After the Cultural Revolution there was a major change in teaching and research activities. Prior to this change, students followed a traditional program of 4 years of study. This has now been reduced to 2 years for agronomy or 3 years for veterinary science, on the assumption that all new students will have spent at least 2 years on a commune. Students may enroll in one of 8 departments: agronomy, forestry, plant protection, horticulture, animal husbandry and/or veterinary science, agricultural engineering, agricultural economics, and hydrology.

The library of this college has over 410,000 volumes. A number of key journals in English are received. Large reading rooms are available for each of the special subject areas, and key reference texts are held on reserve in these rooms.

Students may follow study programs of specialization, including agricultural crops (agronomy), fruit crops, animal husbandry, plant protection, irrigation, water conservation, design and manufacture of agricultural machinery, use and maintenance of agricultural machinery, forestry, and veterinary science.

Before the Cultural Revolution, there were 4,000-5,000 students. However, both students and faculty were sent to various communes and all classes were suspended for 2 or 3 years.

There are 1,476 staff and teachers, including 370 cadres (administrative personnel), 63 professors and associate professors, 221 lecturers (instructors), and 228 assistants. There are 2,000 students. In addition to faculty at the college, various scientists at the nearby academy are invited to present lectures. Also, scientists and professors from other colleges offer lectures and seminars.

As is true for all other organizations, the governing body of the college is called the "Revolutionary Committee." This committee usually includes the president or director, certain administrative officers,

TABLE B-1 Prices of Agricultural Products at a Peking Market, as Posted on August 13, 1975

Vegetables and fruits	Yuan/km[a]	$/pound
Leek, seedlings	0.82	0.37
Leek blossom	0.28	0.13
Broccoli	0.22	0.10
Green onion	0.10	0.05
Eggplant	0.10 (0.12)	0.05
Snapbeans	0.26 (0.38)	0.12
Tomato		
In season	0.14	0.06
Out-of-season	0.34 (0.44)	
Pumpkin	0.10 (0.14)	0.05
Winter melon	0.06 (0.09)	0.03
Cucumber	0.06 (0.18)	0.03
Bitter gourd	0.26 (0.24)	0.12
Sponge gourd	0.10 (0.18)	0.05
Broad beans	0.14	0.07
Fresh soybean	0.14 (0.14)	0.07
Cauliflower	0.14 (0.38)	0.07
Lily flower buds	0.05	0.23
Sea weed (dry shredded)	1.34	0.61
Persimmon	0.18	0.08
Meats and seafoods		
Beef	1.50	0.68
Lamb	1.42	0.64
Pork	1.80	0.82
Chicken (live)	0.93-1.20 (1.60-1.92[b])	0.42-0.54
Duck	2.40 (1.80-2.86)	1.09
Fish (fresh-water)	0.80	0.36
Shrimp (frozen)	3.10	1.41

[a]Prices in parentheses are those reported by the U.S. NAS Plant Studies Group for items in a Shanghai market in September 1974. Many items were priced lower in Peking in 1975 than in Shanghai in 1974. The differences may be due to differences in season and city. No inflation was evident.

[b]The range is due to size difference; large city is more expensive per unit weight.

faculty members, staff, and students. There are usually 10-12 members on the committee. Decisions as to policy matters are subject to final approval by higher administrative authorities at the provincial level.

刘敬修 Liu Ching-hsiu
Chairman
Revolutionary Committee

李志才 Li Chih-ts'ai
Vice-Chairman
Revolutionary Committee
Office; cereal crops diseases research

李振岐 Li Chen-ch'i
Vice-Chairman and Assistant Professor
Plant Protection Department

朱象三 Chu Hsiang-san
Researcher
Shensi Agricultural Research Institute

袁　峰 Yüan Feng
Lecturer,
Plant Protection Department, cotton insect research

路进生 Lu Chin-sheng
Lecturer and Interpreter,
Plant Protection Department, cotton insect research

强彦文 Ch'iang Yen-wen
Lecturer,
Plant Protection Department

万建中 Wan Chien-chung
Interpreter

贾文林 Chia Wen-lin
Interpreter

张克斌 Chang K'e-pin
Cereal Crops Insects Research

马谷芳 Ma Ku-fang
Plant Protection Department,
Fruit Insect Research

孙益知 Sun Yi-chih
Faculty of Horticulture,
Fruit Insect Research

许玉璋 Hsu Yü-chang
Cotton Cultivation Research

Tsaoyang Workers' New Village, Shanghai

The Tsaoyang Workers' New Village, the first urban housing project that
the People's Republic built for Shanghai workers, was constructed in
the city's northern suburbs in 1951. It contains 16,000 households
which hold 70,000 people. Buildings range from 2 to 5 stories and have
running water, electricity, gas, and some plumbing. The village, situ-
ated on 1.5 km^2 of land, incorporates a full range of facilities, including a theater for movies and opera, a "cultural palace," 7 nurseries,
6 kindergartens, 14 primary schools, 8 middle schools, 2 hospitals, 14
cooperative medical stations, department stores, markets, barbershops,
tailor shops, banks, a laundry, a post office, and a photo studio. The
village has a working population of 38,000, 90% of whom work in factories;
3,000 housewives work in 23 small workshops; and 2,500 retired workers
help in hygiene work and teaching. (Work by the retired is uncompensated
since the retired are on 70% pensions.) The village originally was made
up of those living in the poorest housing (including straw-mat houses and
river boats). These people were selected by the factories in the
Tsaoyang area. Now those relatively better off are nevertheless eligible for workers' housing; the decision as to who gets the opportunity
is made by the village's house management organization, operating under
the authority of the Shanghai Housing Bureau.

Shanghai Academy of Agricultural Sciences, Shanghai

The Shanghai Academy of Agricultural Sciences, established in 1960, was
based on an agricultural experimental farm that had been started in 1956.
There are about 1,300 staff members of the Academy, with about 400 of them
classified as scientific researchers. There are 5 research institutes
at the Academy: plant breeding, horticulture, livestock and veterinary
work, agricultural sciences, and soil fertilization. The Institute of
Plant Protection has a staff of about 180; about 90 are scientific technicians and the rest are workers. This institute has 5 hectares of
experimental land.

The Institute of Plant Protection does research in its laboratories
and does a large amount of popularizing of plant protection methods.
To this end, about one-third of the staff is at all times away from the
Academy working at the communication points that are scattered throughout the agricultural communes around Shanghai.

楊阿掌　Yang A-chang
　　　　　Responsible Member
　　　　　Revolutionary Committee

王仁饗　Wang Chen-ts'uan
　　　　　Responsible Member
　　　　　Institute of Soil Fertilizer and Plant Protection

李甫达　Li Fu-ta
　　　　　Researcher
　　　　　Scientific Production Section

彭運祥　P'eng Yün-hsiang
　　　　　Researcher
　　　　　Institute of Soil Fertilizer and Plant Protection

張根乔　Chang Ken-chiao
　　　　　Researcher
　　　　　Institute of Soil Fertilizer and Plant Protection

姚文岳　Yao Wen-yüeh
　　　　　Researcher
　　　　　Institute of Horticulture,
　　　　　Fungus Research

吴印青　Wu Yin-ch'ing
　　　　　Researcher
　　　　　Institute of Horticulture,
　　　　　Fungus Research

趙庆华　Chao Ch'ing-hua
　　　　　Researcher
　　　　　Institute of Crop Breeding and Cultivation

Shanghai Institute of Entomology, Shanghai

When it was established in 1958, the Shanghai Institute of Entomology
had only 20 members, who were devoted almost exclusively to the classi-
fication of insects. It now has 170 members, with 3 research groups
and 3 research departments. The institute was originally part of the
Academia Sinica, but since 1970 it has come under the jurisdiction of
the Shanghai Municipal Revolutionary Committee, with the result that
most of the research is limited to problems at the local Shanghai level.
One of the areas in which the research groups have been working is
insect pheromones.

徐 克　Hsü K'e
　　　　Chairman
　　　　Revolutionary Committee

楊 平 瀾　Yang P'ing-lan
　　　　　Vice-Chairman
　　　　　Revolutionary Committee

魯 华　Lu Hua
　　　　Member
　　　　Revolutionary Committee

祁 云 台　Ch'i Yün-t'ai
　　　　　Member
　　　　　Revolutionary Committee;
　　　　　Member, Pheromone Research Group

石 奇 光　Shih Ch'i-kuang
　　　　　Secretary
　　　　　Scientific Research Department, Production Section

朱 国 凯　Chu Kuo-k'ai
　　　　　Science Worker

邵 柏 祥　Shao Po-hsiang
　　　　　Responsible Member
　　　　　First Research Laboratory

周 振 惠　Chou Chen-hui
　　　　　Science Worker

候 能 俊　Hou Neng-chün
　　　　　Science Worker

彭 辛 午　P'eng Hsing-wu
　　　　　Science Worker

唐 贤 汉　T'ang Hsien-han
　　　　　Head
　　　　　Pheromone Research Group

李 文 谷　Li Wen-ku
　　　　　Member
　　　　　Pheromone Research Group

戴 季 达　Tai Chi-ta
　　　　　Member
　　　　　Pheromone Research Group

尹文英　Yi Wen-ying
　　　　　Interpreter

姜家良　Chiang Chia-liang
　　　　　Interpreter

Changchen (Long March) Commune, Shanghai

Changchen Commune is about 40 minutes by car northwest of downtown Shanghai. The commune was founded in 1958 by combining several advanced agricultural cooperatives. It includes 6,700 households (29,000 people) and has 1,082 hectares of land. The commune is composed of 14 production brigades divided into 108 production teams and is primarily engaged in growing vegetables for the population of Shanghai, although it also raises pigs, chickens, sheep, and corn. Its output of vegetables feeds about half a million people. It supplied the city with 2,000 tons of pork in 1974. Production has risen greatly since 1949-- from 22.5 tons per hectare and 79 varieties of vegetables in 1949 to 98 tons per hectare and 194 varieties. There have been other notable improvements: water conservation and irrigation, especially since 1958; the establishment of factories for repair work and for building tools; the creation of tractor and truck stations; progress in education (the 70% pre-1949 illiteracy rate has been reduced by the founding of four middle schools and nine primary schools); and vast improvement in health care. Each production brigade has its own medical station, and the commune has a small hospital. Cultural control of insects and pest forecasting are measures the commune has adopted to deal with its principal insect problems: the black cutworm, red spiders, and the diamond-back moth.

Nanshang Village Patriotic Hygiene Unit, Shanghai

Nanshang Village is about 40 minutes northwest of downtown Shanghai, in Chia-ting County--the same general area as the Long March Commune. Nanshang has 18,000 people living in over 3,500 households. Facilities include 16 factories, 3 primary and 3 middle schools, over 100 shops, a large park that draws visitors from Shanghai City, and a hospital.

Before 1949, the town's hygiene work was rather backward. It was said that there was no running water (dirty river water was used), and mosquitoes, flies, malaria, and cholera were everywhere. Since that time, mosquitoes and flies have been greatly reduced by mass participation in the campaign to eliminate the "four pests" (bed bugs, mice, flies, and mosquitoes). Doctors and medical personnel have spread the health work message. Removal of garbage and night soil has been organized, and cholera, smallpox, schistosomiasis, and a number of other diseases have been eliminated, we were told.

Those in charge of the health effort included:

唐为梯
    T'ang Wei-t'i
      Responsible Member
      Leading Group

陈德耿
    Ch'en Te-keng
      Officer
      Hygiene Unit Office

张洪顺
    Chang Hung-shun
      Officer
      Hygiene Unit Office

Shanghai Institute of Organic Chemistry, Shanghai

The Shanghai Institute of Organic Chemistry, established in 1950, falls under the dual jurisdiction of the Academia Sinica and the Shanghai Municipal Revolutionary Committee. There were about 1,100 staff members at the institute. The research areas in which the institute is involved are: photosynthesis; artificial single-cell protein synthesis; insect hormones, pheromones, and sex attractants; steriod chemistry and the development of birth control agents; fluorocarbon chemistry; organic and metallic compounds, including boron chemistry; and the chemistry of organic extractants and their applications. There is also some work in physical and biological chemistry as they relate to organic chemistry. A pilot plant is involved in the synthesis of protein from petroleum, and there is an organic analysis laboratory.
    Some of the personnel included:

汪 猷
    Wang You
      Vice-Chairman
      Revolutionary Committee
      Professor

顾立仁
    Ku Li-chen
      Worker
      Organization of Scientific Research

周维善
    Chou Wei-shan
      Worker
      Organization of Scientific Research

吉景顺
    Chi Ching-shun
      Worker
      Fluorine Chemistry

陈毓群
    Ch'en Yü-chün
      Worker
      Insect Hormone Chemistry

许兴妹 Hsu Hsing-mei
        Worker
        Biological Oxidation of Petroleum Laboratory

林国祥 Lin Kuo-hsiang
        Worker
        Insect Information Chemistry

王大琛 Wang Ta-ch'en
        Laboratory Director
        Biological Oxidation of Petroleum Laboratory

黄敬坚 Huang Ching-chien
        Professor
        Nucleic Acids Laboratory

张伟君 Chang Wei-chün
        Professor
        Synthesis of Polypeptide Laboratory

倪 元   Ni Yuan
        Professor
        Steroid Contraceptives Laboratory

Shanghai Children's Palace, Shanghai

Shanghai Children's Palace began in 1953 in an opulent structure that
was once the residence of a British capitalist. The adult who spoke
to us (3 "little red soldiers" also talked to us), known as "Master
Mao," was introduced as having been one of the ex-owner's 50 servants
and a ball boy for the ex-owner's sons when they played tennis. Although 2,000 guest children come daily to enjoy the palace's facilities,
1,000 "little red soldiers" who "Serve the People" are permanently assigned to the palace. They have risen from the 11 district palaces,
which draw from children's homes at street-level. They come to the
palace 1 or 2 times a week as guides or to perform in various cultural,
scientific, or educational activities. (The pianists come every day,
for they cannot practice at home.) Once they have reached middle-school
age, the children are no longer eligible for the palace and its activities.

Discussion with Staff of Shanghai Pesticide Factory

The delegation requested a visit in a pesticide factory in the Shanghai
area. This was not permitted, but 11 responsible staff members from
the factory visited us at the hotel. These persons were very knowledge-
able and eager for information. A very profitable exchange of ideas
ensued. Much of the information is presented elsewhere in this report
in the section on use of insecticides.

This factory, 1 or more than 10 in the Shanghai area, produced more than 20,000 tons of 30 insecticidal chemicals in 1974 and had a staff of 856 persons. The insecticides used in the greatest quantities were trichlorfon, dichlorvos, dimethoate, phosmet, fenitrothion, phosphamidon, and malathion. There was much mutual interest in policies and regulations relating to the use of pesticides in China and the United States, in health and environmental hazards of pesticide use, and in such subjects as insecticide resistance and new types of insecticides. The Chinese were greatly interested in structure-activity relationships of various chemicals and in biodegradable insecticides. Both groups felt the discussions were valuable.

趙　俊
Chao Chün
Technical Responsible Member
Shanghai Agricultural Pesticide Factory

范望达
Fan Wang-ta
Chief-engineer
Shanghai Agricultural Pesticide Factory

楊朝祥
Yang Ch'ao-hsiang
Technician
Shanghai Agricultural Pesticide Factory

韋豪
Wei Hao
Technician
Shanghai Agricultural Pesticide Factory

徐子成
Hsü Tzu-ch'eng
Technician
Shanghai Agricultural Pesticide Factory

张家驹
Chang Chia-chü
Technician
Shanghai Agricultural Pesticide Factory

仇长春
Ch'ou Ch'ang-ch'un
Technician
Shanghai Agricultural Pesticide Factory

王正方
Wang Chen-fang
Technician
Shanghai Agricultural Pesticide Factory

缪晓鹏　　Miu Hsiao-p'eng
　　　　　　Technician
　　　　　　Shanghai Agricultural Pesticide Factory

Tangwan Commune, Shanghai

Tangwan Commune is about half an hour by car southwest of the city of
Shanghai. The commune, formed in September 1958, is divided into 11
production brigades and 120 production teams. It has 23,000 people
in 5,400 households. Of these people, 12,000 work on 1,787 hectares of
commune land, 60% of which is used for grain production; 30% is for
cotton and 10% is for vegetables and herbs, mostly medicinal. Sideline
production includes pigs, chickens, fish, and mushrooms. There are 9
factories, including farm-tool and furniture factories. The commune has
had 13 successive bumper harvests, and the yield per hectare has risen
to 11.575 tons. This is twice the 1958 yield and triple that of 1949.
Cotton production is 905 kg of ginned cotton per hectare, up 152% from
1957 and 302% from 1949. Rape seed production reached 2.123 tons per
hectare, up 116% over 1957 and 372% over 1949. Some 135,400 chickens
were raised in 1974, as well as 1,200,000 fish for food. Mushrooms are
raised in facilities covering 20,100 square meters. Mechanization has
progressed to where 23 tractors and 102 hand tractors are able to plow
95% of the land, and seeding and transplanting operations are 80% mechanized. The 14 primary and 3 middle schools educate about 5,000 students
each year. Medical facilities include a small hospital with 20 doctors
and a staff of 42, a clinic with 3 or 4 "barefoot doctors" in every production brigade, and a "barefoot doctor" in every team. Income per
household has risen to nearly $400 a year, twice that of 1957.

Hunan Institute of Plant Protection, Changsha, Hunan Province

The Hunan Institute of Plant Protection, started in 1952, has about 50
scientific researchers, joined by a smaller number of administrative
staff. The institute is primarily concerned with insect control in
Hunan's rice and citrus crops. To this end, the institute has experimental plots in both crops--about 20 hectares in rice and 35 in citrus
fruits. Research is also done in breeding rice for pest resistance.
　　Those at the institute included:

雷慧志　　Lei Hui-chih
　　　　　　Group Leader
　　　　　　Insect Section

陈长铭　　Ch'en Ch'ang-ming
　　　　　　Researcher

陈永年　Ch'en Yüng-nien
　　　　　Technician

谢建文　Hsieh Chien-wen
　　　　　Director

吴德喜　Wu Te-hsi
　　　　　Technician

Foshan City Patriotic Hygiene Movement Unit, Foshan City, Kwangtung Province

Foshan City, a half hour west of Kwangchow by car, is an old handicrafts center with a 1,000-year history. In modern times, the city prospered through its textile, foundry, and porcelain industries. Since 1949, heavy industry has been brought into the area with the construction of steel, chemical, and industrial machinery plants. Foshan has 230,000 people and is divided into 4 townships. In hygiene work, the former backwardness of Foshan has been altered by a mass effort to change the environmental outlook, substituting hygienic habits for old superstitions, and by carrying out disease-prevention work. Much has been done to institute proper treatment of sewage, waste, and rubbish; one facility even uses marsh gas to generate electricity. We were shown an area consisting of 390 households and 4,000 people. Alleys are no more than a meter wide. It has been completely cleaned and has three pipelines for sewage, waste water, and fresh water running to and from the houses.

许正其　Hsü Cheng-ch'i
　　　　　Responsible Member
　　　　　Foshan City Reception Office

陈　伟　Ch'en Wei
　　　　　Responsible Member
　　　　　Office of Foshan City Patriotic Hygiene Movement
　　　　　　Committee

Hsinhua Production Brigade, Shun-te County, Kwangtung Province

Hsinhua Production Brigade is about a 2-hour drive northwest of Kwangchow. Primarily engaged in sericulture, it also raises fish and sugarcane. Last year the brigade had its best silk-production year ever, with an output of 305 jin per mu and 354 mu of mulberry trees. This production brigade initiated the use of juvenile hormones in the

production of silk in 1963. At that point the highest production
was 241 jin per mu.

李国印　Li Kuo-yin
　　　　　Vice-Chairman
　　　　　Shun-te County Revolutionary Committee

胡炳洪　Hu Ping-hung
　　　　　Vice-Director
　　　　　Shun-te County Science and Technology Bureau

黄自然　Huang Tzu-chan
　　　　　Technician
　　　　　Shun-te County Science and Technology Bureau

楊赞喜　Yang Tsan-hsi
　　　　　Technician
　　　　　Shun-te County Hormone Experiment Factory

Tahsia (Big Sand) Commune, Sse-hui County, Kwangtung Province

Tahsia Commune is located along the lower reaches of the West River
(Hsi Chiang) in Sse-hui County, about 2-1/2 hours west of Kwangchow.
The commune has 7,850 families (31,400 people) divided into 8 agricultural production brigades and 1 fishing brigade, which were subdivided
into 214 production teams. The commune had 60,000 mu of grainland and
15,000 mu in cash crops (sugar, jute, soybean). The rice production
of the commune has been as follows:

Year	Annual Rice Production (in millions of jin)
Before Liberation (1948?)	7
1957	22
1966	42
1974	52

　　Insect control work is done by 1,500 plant-protection technicians
and workers and by groups organized in each of the production brigades.
The technicians do their research work and observations in cooperation
with the students and teachers of Chungshan University so that they have
a practical understanding of the university's field tests. Cultural
control is also widely applied, with capital construction work used to
change breeding sources, level land, and get rid of stagnant, trash-
filled ditches. High-quality seed is used, and land is plowed in the
winter to kill the overwintering insects. The commune has its own small
B.t. factory, uses ducks to eat insect pests in the rice paddies, and

rears *Trichogramma*. As a result of these measures, the insect pest
population has decreased greatly, the use of insecticides is much reduced (from 75,000 kg in 1972 on the early crop and 34,000 kg in 1974
to 6,700 kg thus far in 1975).

袁恩耀 Yüan En-yao
Responsible Member
Foreign Affairs, Sse-hui County

古德祥 Ku Te-hsiang
Technician

罗玉钏 Luo Yu-ch'uan
Student
Chungshan University

麦保祥 Mai Pao-hsiang
Vice-Chairman
Revolutionary Committee

陈国麟 Ch'en Kuo-lin
Responsible Member
Foreign Affairs, Shaoshe Area Revolutionary Committee

八碧金 Pa Pi-chin
Responsible Member
Foreign Affairs, Shaoshe Area Revolutionary Committee

冯剑明 Feng Chien-ming
Responsible Member
Shaoshe Area Revolutionary Committee

Chungshan University (Sun Yat-sen University), Kwangchow, Kwangtung Province

Sun Yat-sen established Kwangtung University in 1924, and after his death
it was renamed to commemorate his democratic revolutionary spirit. In
1931, the newly renamed university was moved from downtown Kwangchow
to what was then part of the campus of Ling-nan University, a site that
it presently occupies. It has grown from a small university with departments specializing in literature and arts, science and technology,
and foreign languages to the major university in southern China, with
departments in Chinese, history, philosophy, economics, foreign languages, mathematics, physics, chemistry, metal sciences, geography, and
biology. Within the biology department are 5 specialties: biological
chemistry and microbiology, entomology, medicinal plants, zoology, and
plant physiology/genetics of plants.

There are 2,600 students in 3-year programs, with a projected 5,000
students by 1980. This compares with 4,700 students before the Great

Proletarian Cultural Revolution. The students spend a third of their time on farms, communes, and factories. There are about 2,000 teachers and staff members at the university; about half of these are in the instructional part of the program.

The biology department has two research institutes, one in ecological entomology and the other in hydrobiology. The department has attached to it a pilot plant and an experimental farm specializing in genetic breeding and pond culture. There are about 300 students in the department, 40 of whom are in their third year. The department has about 200 teachers and staff. The laboratories are equipped with two electronic microscopes--a Chinese-manufactured one used for teaching and a Japanese one used for research.

浦蛰龙   P'u Che-lung
       Professor
       Biology Department

刘翠英   Liu Ts'ui-ying
       Professor
       Biology Department

刘复生   Liu Fu-sheng
       Lecturer
       Biology Department, Entomological Speciality

黄献之   Huang Hsien-chih
       Student
       Biology Department, Entomological Speciality

贺衡萍   He Heng-p'ing
       Student
       Biology Department, Entomological Speciality

魏聪桂   Wei Ts'ung-kui
       Responsible Member
       Educational Revolution Section

龙康候   Lung K'ang-hou
       Professor
       Chemistry Department

齐雨藻   Ch'i Yü-tsao
       Vice-Chairman
       Biology Department

Kwangtung Institute of Entomology, Kwangchow, Kwangtung Province

This institute was originated in 1958 during the Great Leap Forward as the South China Institute of Entomology. During the Great Proletarian Cultural Revolution, the name of the organization was changed to the Kwangtung Institute of Entomology, reflecting a more localized area of research responsibility and a more decentralized level of administrative control--the provincial revolutionary committee.

There are 130 staff members and research workers at the institute, representing persons trained in universities, those who have graduated from middle and technical schools, and technicians who have been trained since Liberation in factories and laboratories. There are 4 laboratories: biological control research, termite control research, zoological research, and entomological resources research.

Many of the staff members spend time working at commune-level research and popularization points. Tahsia (Big Sand) Commune is a good example of this working relationship. Many of the members of the research staff of the institute have been involved with a broad range of insect control methods on the commune's rice, sugarcane, jute, and soybeans.

郎吉成　Lang Chi-ch'eng
　　　　　Vice-Chairman

陈铨晨　Ch'en Ch'üan-wei
　　　　　Head
　　　　　Production Section

麦秀慧　Mai Hsiu-hui
　　　　　Head
　　　　　Citrus Insects Biological Control Section

王鼎祥　Wang Ting-hsiang
　　　　　Researcher
　　　　　Lac Insect Control Section

刘南欣　Liu Nan-hsin
　　　　　Research Worker
　　　　　Beneficial Insects

谢中能　Hsieh Chung-neng
　　　　　Technician
　　　　　Biological Control Laboratory

秦谈亮　Ch'in T'an-liang
　　　　　Responsible Member
　　　　　Zoology Laboratory

李丽英　Li Li-ying
　　　　　Researcher
　　　　　Rice Insect Integrated Control

<u>Visit to Botanical Garden in Kwangtung</u>

The Kwangtung Botanical Garden is affiliated with the Kwangtung Institute of Botany, and two members of the delegation were able to visit it,
although it was closed for renovation. This is an outstanding botanical
garden and has over 3,000 species of plants in an attractive area of
10,000 mu, about 20 km from Kwangchou. The garden was only begun in
1958 and now has lakes, paths, shade gardens, with over 200 species of
medicinal plants, and a very attractive visitor center. There are outstanding collections of orchids, of bamboo with over 100 species, palms,
water plants, and so on. The director and assistant director were very
well trained and informed us that the garden had cooperative agreements
with 40 countries to exchange plant collections. The garden was described as having 3 broad purposes: display of plants, advancement of
economic botany, and development of better varieties of certain plants.

郭俊彦　Kuo chün-yen
　　　　Director

Appendix C

COMMON PESTS (INSECTS AND MITES) FOUND

IN CHINA

This list is based on lectures and demonstrations presented during the visit of the American Insect Control Delegation to the People's Republic of China in August 1975 and on information obtained from various Chinese publications. The common names of pests were translated from Chinese by Dr. Po-Yung Lu, Dr. Chang-Shyan Chang, and Mr. Lam Au. Names in parentheses are U.S. common names, approved by the Entomological Society of America, or alternative translations of Chinese names.

APPLE

Acleris crocopepla (Meyrick)	leafroller
Acronycta increta Buller	peach striped noctuid
Adoxophyes congruana Wallcer	apple leafroller
Adoxophyes orana Fischer von Röslerstams	leafroller
Agrilus mali Matsumura	apple tree buprestid
Anomala corpulenta Motsch	brown scarabaeid
Anoplophora chinensis Forster	star longicorn beetle
Aphis pomi De Geer	apple aphid
Apriona germari Hope	mulberry longicorn beetle
Archips asiaticus (Walsingham)	leafroller
Archips breviplicanus (Walsingham)	leafroller
Archips xylosteanus (Linnaeus)	leafroller
Aspidiotus perniciosus Comstock	round pear scale (San Jose scale)
Autoserica japonica Motsch	red velvet scarabaeid
Bacchisia fortunei Thomas	
Bryobia praetiosa Koch	clover mite
Carposina niponensis	
Carposina sasakii Matsumura	small peach fruit borer (peach fruit moth)
Choristoneura lonicellana (Walsingham)	leafroller
Conopia hector Buller	apple aegeriid
Eriosoma lanigerum (Hausmann)	woolly apple aphid
Gastropacha quercifolia Linnaeus	apple lasiocampid
Grapholitha inopinata Heinrich	small apple fruit borer (Manchurian fruit moth)

189

Grapholitha molesta (Busck)	small pear fruit borer (oriental fruit moth)
Holotricha diomphalia Bates	Korean scarabaeid
Holotricha serobiculata Brenske	large black scarabaeid
Hypocana subsature Guenee	apple noctuid
Hyponomeuta padellus Zeller	
Illiberis pruni Dyar	leafroller
Leucoptera scitella	leafroller
Linda fraterna (Chevr)	apple longicorn beetle
Lithocolletis ringoniella	leaf miner
Lithocolletis triflorella Reger	golden-striped leaf miner (apple leaf miner)
Malacosoma neustria testacea Motschulky	heaven tent caterpillar
Myzus malisulcatus Matsumura	apple gall aphid (apple leaf curling aphid)
Obera japonica Thunberg	apple twig longicorn beetle
Pandemis heparana (Schiffermüller and Denis)	brown apple leafroller (apple brown tortrix)
Pandemis ribeana Hübner	
Panonychus ulmi Koch	apple red mite (European red mite)
Phalera flavescens Bremer and Grey	
Rhopobota naevana Hübner	leafroller (black headed fireworm)
Serica orientalis Motsch	black velvet scarabaeid
Spilonota lechriaspis Meyrick	leafroller
Spilonota prognathara Snoll	white apple borer
Stephanitis nashi Esaki et Takeya	pear stink bug
Telphusa chloroderces Meyrick	black star moth
Tetranychus viennensis Zacher	sweet cherry spider mite

BANANA

Calandra sordide Germar	banana weevil
Erionota thorax Linnaeus	banana leafroller
Odoiporus sp.	double black banded weevil

BEANS

Agrotis ypsilon Rottemberg	(black cutworm)
Aphis medicaginis Kock	leaf aphid (cowpea aphid)
Aphis persie	
Aphis tabacci Lindeman	onion thrips
Clanis bilineata Walker	bean hawk moth
Epicauta gorhami Marseul	bean blister beetle
Phytometra agnata Staudinger	silver-striped bean noctuid
Phytomyza atricornis Meigen	leaf miner
Prodenia litura Fabricius	tobacco cutworm

CABBAGE, TURNIP, AND RELATED VEGETABLES

Acrida chinensis Westw.	Chinese grasshopper
Acrida turrita Linnaeus	grasshopper
Agrotis ypsilon Rottemberg	(black cutworm)
Athalia rosae Linnaeus	vegetable sawfly
Atractomorpha sinensis Bolivar	grasshopper
Aulacophora femoralis Motschulsky	cucurbit leaf beetle
Autographa nigrisima Walker	looper
Barathra brassicae Linnaeus	cabbage armyworm
Brevicoryne brassicae (Linnaeus)	cabbage aphid
Cifuna locuple Walker	tussock moth
Colaphellus bouringi Baly	leaf beetle
Diacrisia sucarnea	woolly bear
Epilachna sparsa orientalis Dieke	sour juice lady beetle
Epilachna 28-maculata Motschulsky	28-spotted lady beetle
Hellula undalis Fabricius	vegetable borer
Hylemya platura Meigen	melon seed fly
Laphygma exigua Hübner	sugarbeet noctuid
Pegomyia hyosciami (Panzer)	spinach leaf miner
Phaedon brassicae Baly	small leaf beetle
Phyllotreta humilis Weise	flea beetle
Phyllotreta rectilineata Chen	flea beetle
Phyllotreta vittata Fabricius	striped flea beetle
Phyllotreta vittula Redt	flea beetle
Phytomyza atricornis Meigen	garden pea leaf miner
Pieris rapae Linnaeus	common cabbage worm
Plusia agnata Staudinger	silver noctuid
Plutella maculipennis (Curtis)	diamondback moth
Prodenia litura Fabricius	tobacco cutworm
Rhopalosiphum pseudobrassicae (Davis)	turnip aphid
Trichoplusia agnata	looper

CHESTNUT

Cinalathura folial (Theobald)	big aphid
Cinara pinea Mordwike	big pine aphid
Curculio davidi Fairmaire	chestnut weevil
Dryocosmus kuriphilus Yasurnatus	chestnut gall wasp
Gryllorhynobites ursulus Roelofs	tree-cutting weevil
Pterochlorus tropicalis Van der Grod	big chestnut aphid

CITRUS

Adoxophyes privitana	leafroller
Agrilus citri Matsumura	flatheaded citrus borer
Aleurocanthus spiniferus Quaintance	black spiny whitefly
Aleurolobus marlatti Quaintance	

Anoplophora chinensis (Forster)	white-spotten longicorn
Aonidiella aurantii Maskell	red scale
	(California red scale)
Aphis citricidus Kirkaldy	citrus aphid
Cacoecia asiatica Walker	yellow-tailed leafroller
	(citrus leafroller)
Ceroplastes rubens Maskell	red wax scale
Chrysomphalus ficus Ashmead	brown round scale
Clania minuscula Butler	tea bagworm
Clitea metallica	citrus leaf beetle
Contarinia sp.	citrus bud maggot
Contarinia citri Barnes	citrus flower-bud midge
Homona coffearia	leafroller
Icerya purchasi Maskell	cottony cushion scale
Lepidosaphes beckii (Newman)	(purple scale)
Nadezhdiella cantori	brown longicorn beetle
Panonychus citri (McGregor)	citrus red mite
Papillio xuthus Linnaeus	spring swallowtail
	(smaller citrus dog)
Parasa consocia Walker	green cochlid
Parlatoria pergandii Comstock	chaff scale
Parlatoria zizyphys Lucas	
Phyllocnistis citrella Stainton	citrus leaf miner
Phyllocoptruta oleivora (Ashmead)	citrus rust mite
Prontaspis yanonensis Kuwana	arrowhead scale
Pseudococcus citri Risso	citrus mealybug
Schizotetranychus sexmaculatus	mite
	(six-spotted mite)
Tetradacus citri Chen	large citrus fruit fly
Unaspis yanonensis Kuwana	

CORN AND SORGHUM

Acrida chinensis Westw.	Chinese grasshopper
Anticyra combusta Walker	cerurid
Aphis sacchari Zehntner	sorghum aphid
Atractomorpha lata (Motschulsky)	long-faced locust
Calliptamus abbreviatus Ikonn	short back-winged locust
Chaetocnema ingenus Baly	grain stem flea beetle
Chilotraea infuscatellus Snellen	sorghum borer
Corizus hyalinus Fabricius	sorghum stink bug
	(hyaline grass bug)
Cryptoblabes sp.	earworm
Diatraea shariinesis Eguchi	sorghum grey borer
Diatraea venosata Walker	striped sorghum borer
Euprepocnemis shirakii I. Bol.	black-backed locust
Locusta migratoria manilensis Meyen	oriental migratory locust
Mampava bipunctella Ragonot	sorghum earworm
Oedaleus infernalis Saussure	yellow-necked locust

Ostrinia (Pyrausta) nubilalis (Hübner)	corn borer (European corn borer)
Pseudaletia separata Walker	armyworm
Rhopalosiphum prunifoliae Fitch	corn aphid
Thrips chinense	thrips
Lema tristis Herbst	grain beetle (yellow-legged lema)

COTTON

Adelphocoris fasciaticollis Reuter	three-spotted blind stink bug
Adelphocoris lineolatus Goeza	alfalfa plant bug
Adelphocoris suturalis Jakovlev	black-striped plant bug
Adelphocoris taeniophorus Reuter	three-spotted plant bug
Adoxophyes orana Fischer von Roslerstamm	cotton leafroller
Agrotis takionis Butler	cutworm
Agrotis ypsilon Rottemberg	black cutworm
Anomis fimbriago Steph	noctuid
Anomis flava Fabricius	cotton looper (cotton leaf caterpillar)
Anomis fulvida (Guenee)	noctuiid
Anomis mesogonia Walker	noctuiid
Aphis gossypii Glover	cotton aphid
Archips rosaceana Harris	leafroller
Ascotis selenaria Schiffenmüller et Denis	big cotton looper
Calomyclerus sp.	small grey weevil
Chlamydatus pullus Reuter	small blind stink bug
Chlorita (Empoasca) biguttula Shiraki	cotton leaf hopper
Chondracris rosea Degeer	large green grasshopper
Cicadula fasciifrons Stoll	cotton leafhopper
Cocytodes coerulea	sphingid
Creontiades gossypii Hsiao	plant bug
Deraecoris punctulatus Fall	black-spotted blind stink bug
Diacrisia lubricipeda Linnaeus	yellow arctiid
Diacrisia subcarnia Walker	red arctiid
Dysdercus cingulatus Fabricius	plant bug
Earias cupreoviridis Walker	cotton green moth
Earias insulana Boisduval	banded cotton bollworm (spiny bollworm)
Earias pudician Staudinger	one-spotted diamond bollworm
Earias tabia Stoll	green-striped bollworm
Euxoa (Agrotis) segetum Schiff.	cutworm
Heliothis armigera (Hübner)	cotton bollworm
Heliothis asfulta Guenee	diamond bollworm
Heliothis dipsacea (Fabricius)	clover noctuid
Heliothis peltigera (Schiff.)	big cotton bollworm
Hylemya platura Meigen	seed maggot
Lygus lucorum Mayer-Dur	green plant bug
Lygus pratensis Linnaeus	pasture grass blind stink bug
Meliclepteriq scutosa (Schiff.)	wide-necked noctuid

Neomargarodes sp.	cotton pearl scale
Pareba vesta Fabricius	butterfly
Pectinophora gossypiella (Saunders)	pink bollworm
Prodenia litura Fabricius	armyworm
Pyrameis indica Herbst	butterfly
Sylepta derogata Fabricius	large cotton leafroller
Tetranychus bimaculatus Harvey	two-spotted spider mite
Tetranychus telarius (Linnaeus)	cotton red spider mite
Thrips tabaci Lindeman	cotton thrips (onion thrips)
Zeuzera coffeae Neitner	zebra-striped moth

DATE

Apochemia sp.	date looper
Carposina niponensis Walsingham	small peach borer
Ceroplastes japonicus Gory	turtle wax scale
Cerostoma sasakii Matsumura	date sticky-worm
Chalcophora japonica Gory	japanese borer
Chrysobothris succedanea Saunders	six-spotted borer
Lecanium excrescens Ferris	date bald scale
Monema flavescens (Walker)	yellow spiney-moth
Narosa edoensis Kawoda	small spiney-moth
Parasa consocia Walker	green spiney-moth
Scythropus yasamatsui Kono et Morimoto	small grey weevil

EGGPLANT

Acrida sp.	grasshopper
Agrotis ypsilon Rottemberg	(black cutworm)
Atractomorpha sp.	grasshopper
Epilachna sparsa orientalis Dicke	

GRAPE

Dactylosphaera vitifolii Fitch	grape root aphid
Empoasca sp.	leaf hopper
Eriophyes vitis Pagenst	grape mite
Erythroneura apicalis Nawa	grape leafhopper
Grapholitha molesta (Busck)	oriental fruit moth
Oides decempunctata Billberg	leaf beetle
Paranthrene vegale Butler	grape aegeriid
Parthenolecanium orientalis Borche	flat scale
Phylloxera vitifoliae (Fitch)	grape leaf louse (grape phylloxera)
Sciapteron regale Butler	grape clearwing moth
Xylotrechus pyrrhoderus Bates	grape borer

HEMP

Mordellistena cannabisi Matsumura	flea beetle
Paraglenea fortunei Saunders	cerambycid
Phalonia epilinana Linnaeus	microlepidoptera
Psylliodes attenuata Koch	hemp flea beetle
Thyestilla gebleri Faldermann	hemp longicorn beetle

LITCHI

Acrocercops cramerella Snell	litchi bark miner
Aristobia testudo Voet	longicorn beetle
Cacoecia asiatica Walker	yellow leafroller
Cerace stipatana Walker	leafroller
Deudorix epijarbas Moore	litchi gray butterfly
Tessaratoma papillosa Drury	litchi stink bug

MAN AND ANIMALS

Culicidae

Aedes aegypti (Linnaeus)
Anopheles hyrcanus sinensis Wied.
Anopheles hyrcanus lesteri
Aedes albopictus Skuse
Culex bitaeniorhynchus Giles
Culex fuscanus Wied.
Culex pipiens pallens Coq
Culex tritaeniorhynchus Giles
Culex vagans Wied.
Culex vishnui Theobold
Culex vorax Edwards

Calliphoridae

Achaetandrus rufifascies
Aldrichina grahami Ald.
Bengalia varicolor Fabricius
Calliphora vicina R.
Calliphora vomitoria Linnaeus
Chrysomya bezzia Vill.
Chrysomya megacephala Fabricius
Chrysomya pinguis Walker
Hemipyrellia ligurriens Weidemann
Idiella tripartita Bigor
Lucilia cuprina Weidemann
Lucilia illustris Meigen
Lucilia porphyria Walker

Lucilia sericata Meigen
Phormia regina Meigen
Polleniopsis mongolica Seguy
Protophormia terraenovae R-D
Triceratopyra calliphoroides Rohd.

Muscidae

Atherogonia undiseta megaloba
Fannia canicularis Linnaeus
Fannia priscea
Graphomya rutitibia Stein
Hydrotaea dentipes
Lispe orientalis
Lyperosia exigua Meigen
Mesembrina decipiens Lu
Morellia sinensis Ouchi
Musca amita Hennig
Musca bezzii Patton and Craig
Musca canducens Walker
Musca domestica vicina
Musca hervei Vill.
Musca sorbens sorbens
Muscina angustifrons Lu
Ophyra chalcogaster
Ophyra leucostoma Weid.
Orthiella cocrulea Weid.
Orthiella violacia Macq.
Pyrellia cadaverina
Stomoxyx calcitrans Linnaeus

Sarcophagidae

Bellieria melanara Meigen
Beraea haemorrhoidalis Fallen
Boettcherisca peregrina R-D
Metopix leucocephala Ross
Parasarcophaga crassipalpis Maeq
Pierretia ugamskii Rohd.
Ravinia striata Fabricius
Wohlfahrtia cheni Rohd.

Blattidae

Blattella germanica (Linnaeus)
Periplaneta americana (Linnaeus)
Periplaneta emarginata

ONION

Helix grahainum Linnaeus	snail
Prodenia litura Fabricius	tobacco cutworm
Thrips tabacci Lindeman	onion thrips

PEACH, APRICOT, PLUM

Aromia bungii Falderman	red-necked longicorn
Chlorita flavescens Fabricius	peach leaf hopper
Cicadella viridis Linnaeus	green leaf hopper
Dermaleipa juno (Dalman)	hairy noctuid
Dichocrocis punctiferalis Guenee	peach borer (peach pyralid moth)
Didesmococcus koreanus Borchs	Korean scale
Emposca flavescens Fabricius	smaller green leaf hopper
Eulecanium prunastri Fonsee	apricot scale
Grapholitha molesta Busck	small pear fruit borer (oriental fruit moth)
Herdonia osacesalis Walker	pomegranate thysid
Hopocampa sp.	plum sawfly
Hyalopterus arundinis Fabricius	large-tailed peach aphid
Lecanium kunoense Kuwana	globular peach scale (ume globose scale)
Lyonetia clerkella Linnaeus	peach leaf miner
Marumba gaschkewitschii ecephron Boisduval	peach hornworm
Myzus momonis Matsumura	peach gall aphid
Myzus persicae (Sulzer)	(green peach aphid)
Odonestis pruni Linnaeus	plum lasiocampid
Panonychus ulmi (Koch)	European red mite
Pseudaulacaspis pentagona Targ	mulberry scale
Rhynchites bacchus Linnaeus	small peach weevil (peach curculio)
Tetranychus viennensis Zacher	sweet cherry spider mite
Tischeria sp.	peach leaf miner

PEAR

Acrocercops astourota Meyrick	pear gracilarid
Actias selene Hübner	green saturnid
Anthonomus pomorum Linnaeus	pear blossom weevil
Cinacium iakusuiense Kishi	
Cryptotympana atrata (Fabricius)	black cicada
Grapholitha molesta Busck	small pear fruit borer (oriental fruit moth)
Halyomorpha picus (Fabricius)	tree stink bug
Hoplocampa pyricola Rohwer	pear fruit sawfly

Illiberis pruni Dyar — pear spotted caterpillar (pear leaf worm)

Lampra limbata Gebl — gold buprestid
Lampra sp. — green peach buprestid
Metzenceria sp. — pear gelechiid
Nephoteryx pirivorella Matsumura — large pear fruit moth
Paropsides duodecimpustulata Gebler — pear leaf beetle
Psylla pyricola Forster — pear sucker (pear psylla)

Rhynchites corneanus Kono — pear leaf weevil
Stephanitis ambigua Horvath — pear lace bug
Toxoptera piricola Matsumura — pear aphid
Urochela luteovaria Distant — pear stink bug

PERSIMMON

Acanthococcus kaki Kuwana — woolly persimmon scale
Culcula panterinaria Bremer et Grey — looper
Drosicha corpulenta (Kuwana) — grass shoe scale
Erythroneura mori Mats — blood spot leaf hopper
Kakivoria flavofasciatus Nagano — persimmon budworm
Lymantria dispar Linnaeus — persimmon caterpillar (gypsy moth)

Percnia giraffata Guenee — spotted persimmon looper
Phenacoccus pergandei Cockerell — persimmon long woolly scale
Pseudococcus citri Risso — persimmon mealy bug (citrus mealy bug)

PINEAPPLE

Dysmicoccus brevipes Cockerell — pineapple mealy bug
Liogryllus bimaculatus Degger — sugarcane black cricket

POTATO

Epilachna 28-maculata Motschulsky — 28-spotted lady beetle
Gnorimoschema operculella Zeller — potato tuberworm
Ophiomyia sp. — beanstalk miner

RICE

Ancylolomiae japonica Zeller — rice tube moth
Chilo suppressalis Walker — rice stem borer
Chilotraea awricilia Dudgeon — Taiwan rice borer
Chironomus sp. — rice red maggot
Cicadella viridis — big green leaf hopper

Cicadula fascifrons Stal	two-spotted leaf hopper
Cletus trigonus Thunberg	slender rice bug
Cnaphalocrocis medinalis Guenee	grass leafroller
Delphacades striatella Fallen	grey rice leaf hopper
Deltocephalus dorsalis Motschulsky	zigzag-striped leaf hopper
Dolycoris baccarum Linnaeus	delicate haired bug
	(sloe bug)
Echinocnemus squameus Billberg	rice weevil
Empoasca flavescens Fabricius	small green leaf hopper
Empoasca subrufa Melichar	leaf hopper
Haplothrips aculeatus Fabricius	rice stem thrips
Hispa armigera Olivier	rice armored beetle
Hydrellia griseola Fallen	midge
Laodelphax striatella Fallen	plant hopper
Lema oryzae Kuwayama	rice leaf beetle
Leptocorixa varicornis Fabricius	Corbett rice bug
Mycalesis gotama Moore	rice eye-butterfly
Megarrhamphus hactatus Fabricius	spindle-shaped stink bug
Menida histrio Fabricius	small red stink bug
Mythimna separata (Walker)	oriental armyworm
Naranga aenescens Moore	rice green caterpillar
Nephotettix bipunctatus Fabricius	seven-spotted black-tailed
	leafhopper
Nephotettix cincticeps Uhler	black-tailed leafhopper
	(green rice leafhopper)
Nezara viridula (Linnaeus)	rice green stink bug
	(southern green stink bug)
Nephe elongata Dallas	white-bordered stink bug
Nilaparvata lugens Stäl	brown-backed planthopper
Niloparvata oryzae Matsumura	brown planthopper
Nymphula depunctalis Guenee	three-spotted rice borer
Oxya chinensis Thunberg	rice grasshopper
Pachydiplosis oryzae Wood-Mason	rice gall midge
Parnara guttata Bremer et Grey	rice leafroller
	(rice plant skipper)
Schoenobius sp.	brown-bordered rice borer
Scotinophara lurida Burmeister	black rice stink bug
Sesamia calamistris Hampson	large rice borer
Sesamia inferens Walker	purplish stem borer
Sogata furcifera Horvath	white-backed planthopper
Spodoptera mauritia Boisduval	lawn armyworm
Susumia exigua Butler	striped rice leafroller
Tetroda histeroides Fabricius	yellow-striped stink bug
Thrips oryzae Williams	rice thrips
Tryporyza incertulas Walker	rice paddy borer

SOIL INSECTS

Agriotes fusicollis Miwa
Agrotis c-nigrum (L.)

Agrotis canescens (Butler)
Agrotis exusta (Butl.)
Agrotis fusciollis Miwa
Agrotis tokionis Butl.
Agrotis ypsilon (Rott.)
Anomala axoleta Faldermann
Anomala corpulenta Mots
Dorcus titanus Boisduval
Dorysthenes hydropicus (Pascoe)
Euxoa segetum (Schif.)
Gonocephalum mongolicum Reitter
Gryllotalpa africana Palisot de Beauvois
Gryllotalpa unispina Saussure
Gryllus testaceus Walker
Holotrichia diomphalia Bates
Holotrichia parallela Mots.
Holotrichia titanus Reitter
Holotrichia trichophora Fairm
Hylemya antiqua Meigen
Hylemya floralis (Fallen)
Hylemya pilipyga Velleneure
Hylemya platura (Meigen)
Lepyrus japonica Roelofs
Licola brevitarsis Lewis
Loxoblemuus doenitzi Stein
Melanotus caudex Lewis
Melontha sp.
Opatrum subaratum Faldermann
Phyllopertha pubicollis Waterhouse
Pleonomus canaliculatus Faldermann
Popillia japonica Newm.
Serica orientalis Mots.
Stiboropus flavidus Signoret
Trematodes tenebroides Pallas

SOYBEAN

Agromyza phaseoli Coquillett	bean stalk fly
Amyna octo Guenee	soybean small noctuid
Aphis glycines Matsumura	soybean aphid
Argyrogramma agnata Staudinger	silver striped noctuid
Autographa nigrisima Walker	black-spotted silver stripe noctuid
Cauninca annetta Butler	
Cifuna locuple Walker	soybean tussock moth
Clanis bilineata Walker	bean hawkmoth (greenish brown hawkmoth)
Coptosoma punctatissima Montandon	bean stink bug (globular stink bug)
Epicauta gorhami Marseul	bean blister beetle

Etiella zinckenella (Treitschke)   pea-pod borer
                                     (lima-bean pod borer)

Grapholitha glycinivorella Matsumura   soybean pod borer
Heliothis dipsacca Linnaeus   clover noctuid
Herse convolvuli Linnaeus   convolvulus hawkmoth
Lamprosema indicata Fabricius   soybean leafroller
Maruca testulalis Geyer   soybean borer
Monolepta nigrobilineatus Motschulsky two-striped leaf beetle
Nezara antennata Scott   common green stink bug
Phytometra agnata Staudinger   silver-striped bean noctuid
                                         (three-spotted plusia)

Popillia japonica Newman   Japanese beetle
Tetranychus bimaculatus Harvey   soybean red mite
                                    (two-spotted spider mite)

Trichoplusia ni Hübner   powdery spotted noctuid
                                    (cabbage looper)

SQUASH, CUCUMBER

Agrotis ypsilon Rottemberg   (black cutworm)
Aulacophora cattigarensis Weise   (leaf beetle)
Aulacophora femoralis chinensis Weise   cucurbit leaf beetle
Aulacophora nigripennis Motschulsky   (leaf beetle)
Hellula undalis Fabricius   vegetable borer

STORED PRODUCTS

Alphitobius piceus Oliver
Araecerus fasciculatus De Geer
Bruchus chinensis Linnaeus
Bruchus pisorum Linnaeus   pea weevil
Bruchus rufimanus Boheman   broadbean weevil
Callosobruchus maculatus Fabricius   cowpea weevil
Cylas formicarius Fabricius   sweetpotato weevil
Dermestes cadaverinus Fabricius
Gnorimoschema operculella Zeller   potato tuberworm
Lasioderma serricorne Fabricius   cigarette beetle
Necrobia ruficollis Fabricius   redshouldered ham beetle
Necrobia rufipes De Geer   red-legged ham beetle
Oryzaephilus surinamensis Linnaeus   sawtoothed grain beetle
Plodia interpunctella Hübner   Indian meal moth
Pyralis farinalis Linnaeus   meal moth
Sitophilus granarius Linnaeus   granary weevil
Sitophilus oryzae Linnaeus   rice weevil
Sitotroga cerealella Olivier   Angoumois grain moth
Tenebrioides mauritanicus Linnaeus   cadelle
Tribolium confusum Duval   confused flour beetle
Trogoderma granarium Everts   Khapra beetle

TERMITIDAE

Capritermes fuscotiablis
C. nitobei
Coptotermes brevis
C. declivis
C. domecticus
C. formosanus (14 provinces)   Formosan subterranean termite
C. nitobei (11 provinces)
Dicuspiditermes garthwaitii
Eurytermes isodentatus
Globitermes audax
Glyptotermes ceylonicus
G. chimpingensis
G. currignathus
G. emersoni
G. formosanus
G. fuseus
G. satsumensis
Homallotermes huanensis
Hospitalatermes luzonensis
Hypotermes sumatrensis
Indotermes hanatus
Kalotermes inamurae
Macrotermes barynei (11 provinces)
Microcaprotermes connectens
M. hsuchiafui
Microcerotermes bugniani
M. burmanicus
Nasutitermes cummunis
N. deltacephalus
N. erectinosus
N. fulvus
N. gardneri
N. grandinasus
N. kinoshitae
N. luzonensis
N. moratus
N. orthonasus
N. parafulvus
N. parvanasutus
N. sinuosus
N. takosogensis
Neotermes koshunensis
Odontotermes angustignathus
O. formosanus (15 provinces)
O. hainanensis
O. yunnanensis
Parrhinotermes khasii
Pericaprotermes semarangi
P. tetraphilus

Procapritermes albipennis
P. mushae
P. sowerbyi
Pseudocapritermes minutus
P. pseudolaetus
Reticulitermes aculahialis (Planifrontotermes) (12 provinces)
R. affinis (Frontotermes)
R. assamensis (Frontotermes)
R. chayuensis (Planifrontotermes)
R. chinensis (Planifrontotermes) (13 provinces)
R. curvatus (Planifrontotermes)
R. flaviceps (Frontotermes)
R. fukiensus (Frontotermes)
R. gaoyaoensis (Planifrontotermes)
R. gingdeensis (Planifrontotermes)
R. grandis (Frontotermes)
R. hainanensis (Planifrontotermes)
R. longicephalus (Frontotermes)
R. speratus (Frontotermes)
Schedorhinotermes magnus
S. tarakanensis
Stylotermes latilabrum
S. minutus
S. sinuensis
S. sinuensis inclinatus
S. sinuensis latipeduculus
Termes majori

SUGAR BEET

Aphis fabae Scop	beet aphid
Barathra brassicae Linnaeus	cabbage armyworm
Bothynoderes punctiventris Germar	common beet weevil
Bourletiella pruinosa Tullberg	garden springtail
Cassida nebulosa Linnaeus	beet tortoise beetle
Chaetocnema chalceola Jacoby	beet beetle
Chromoderes fasciatus Müller	striped weevil
Loxostege sticticalis (Linnaeus)	(beet webworm)
Opatrum sabulosum Linnaeus	(darkling beetle)
Pegomya hyoscyami Panzer	spinach leaf miner
Phyllotreta vittata Fabricius	yellow striped beet beetle
Poecyloscytus cognatus Fieb.	plant bug
Prodenia litura Fabricius	tobacco cutworm
Tanymecus palliatus Fabricius	grey beet weevil

SUGARCANE

Aiolopus tamulus Fabricius	mosaic taper-winged grasshopper (green-striped winged grasshopper)
Alissonotum sp.	black beetle
Anomala corpulenta Molisch	chafer
Argyroploce schistaceana Snellen	yellow borer sugarcane shoot borer
Calliptamus abbreviatus Ikonnikov	short spotted-winged grasshopper
Ceratovacuna lagigera Zehntner	woolly sugarcane aphid
Chilo infuscatellus Snellen	two-spotted sugarcane borer
Epacromius coerulipes Ivan	taper-winged grasshopper
Euprepocnemis shirakii I. Bol.	black-backed grasshopper
Gryllotalpa unispina de Saussure	northern China mole cricket
Gryllotalpa africana Palisot de Beauvois	African mole cricket
Holotricha diomphalia Bates	
Locusta migratoria Linne	migratory grasshopper (Asiatic locust)
Oedaleus infernalis de Saussure	false mormorate grasshopper
Proceras venosatus Walker	spotted borer
Prodenia litura Fabricius	tobacco cutworm
Pseudaletia separata Walker	sticky worm (armyworm)
Sesamia inferens Walker	purplish stem borer
Trematodes tenebroides Paller	scarabaeid
Trionymus sacchari Cockerell	sugarcane mealybug

SWEET POTATO

Brachmia triannuella Herrich-Shaffer	sweet potato leaf folder
Cassida circumdata Herbst	sweet potato beetle
Cylas formicarius Fabricius	small weevil (sweet potato weevil)
Herse convolvuli Linnaeus	sweet potato hornworm
Sympiezomias lewisi Roelofs	large grey weevil

TOBACCO

Chloridea assulta Guenee	oriental tobacco budworm
Dolychoris baccarum (Linnaeus)	spotted-fringe stinkbug
Myzus persicae (Sulzer)	(green peach aphid)

WALNUT

Aeoloscelis sp.	moth
Agrilus sp.	walnut twig borer
Batocera horsfieldi Hope	square spotted longicorn
Dyscerus sp.	walnut root weevil
Sphaerotrypes coimbatorensis Steeb	
Zeuzera coffeae Nietner	zebra-striped moth

WHEAT AND BARLEY

Aceria tulipae	gall mite
Aconsosiphum sp.	aphid
Aelia acuminata Linnaeus	pentatomid
Anomala corpulenta Mots.	scarabaeid
Agriotes fusciollis Miwa	wireworm
Apophilia thalassina Fold	chrysomelid
Blissus pallipes Distant	long wheat grain stink bug (chinch bug)
Cephus pygmaeus Linnaeus	wheat stem wasp
Contarinia tritici Kirby	yellow wheat midge
Delphacades striatella Fallen	plant hopper
Dolerus hordei Rhower	barley sawfly
Dolerus tritici Chu	wheat sawfly
Gryllotalpa africana Palisot de Beauvois	mole cricket (African)
Gryllotalpa unispina	mole cricket
Haplothrips tritici	thrips
Holotrichia diomphalia Bates	scarabaeid
Holotrichia titanis Reitter	scarabaeid
Macrosiphum granarium Kirby	tubular wheat aphid (Japanese grain aphid)
Melanotus caudex Lewis	wireworm
Meromyza saltatrix Linnaeus	wheat stem fly
Mythimna separata (Walker)	oriental armyworm
Nanna truncata Fan	tassel fly
Ochsenheimeria sp.	wheat stem borer
Oscinella pusilla	maggot
Pachynematus sp.	sawfly
Penthaleus sp.	round wheat mite (winter grain mite)
Pentodon patrualis Frivalsky	scarabaeid
Petrobia latens (Müller)	long-legged wheat mite (brown wheat mite)
Pleonomus calaliculatus Fold	wireworm
Rhopalosiphum maidis	aphid (corn leaf aphid)
Rhopalosiphum padi Linnaeus	aphid
Rhopalosiphum prunifoliae Fitch	barley aphid
Simotactix striatus	leafhopper

Sitodiplosis mosellana (Géhin)	red sucking fly (wheat midge)
Toxoptera graminum (Rhondani)	greenbug
Trematodes tenebroides Pallas	scarabaeid
Tribolium confusum Jacquelin du Val	confused flour beetle

Appendix D

BOOKS ON AGRICULTURE AND ENTOMOLOGY

These recent books on general agriculture, general entomology, and insect control were collected by the delegation during their trip. Those
marked with an asterisk were presented by Chinese scientists.

Academia Sinica Institute of Experimental Biology. 1966. Everybody can
raise oak silkworm. Scientific Press, Peking. 34 pp. plus one color
plat.

Academia Sinica Lac Survey Team. 1972. Host plants of the lac insect.
Includes glossary of scientific names (in Latin) and Chinese names
of plants. Agricultural Press, Peking. 123 pp.

Agronomy Bureau, Ministry of Agriculture and Forestry. 1975. Handbook
of Agricultural techniques for state farms. Shanghai People's Press,
Shanghai. 632 pp.

Anhwei Province Agriculture Bureau. 1966. Rice borer control. Anhwei
People's Press, Anhwie. 39 pp.

Anonymous. 1970. Control of rice diseases. Chemical Industry Press,
Peking. 56 pp.

Anonymous. 1971. Standards of agricultural chemicals. 1971. Technical Standards Press, Peking. 174 pp.

Chang Kwang-hseh and Wang Ling-yao. 1974. Pictures of cotton insects.
Includes indexes to scientific names (in Latin) and Chinese names of
insects. Scientific Press, Peking. 93 pp. plus 37 color plates.

Chekiang Academy of Agricultural Sciences, Institute of Plant Protection.
1975. Virus diseases of rice. Includes English names of diseases and
scientific names (in Latin) of host plants and insect vectors. Agricultural Press, Peking. 104 pp. plus 5 color plates.

Chinese Academy of Agriculture, Institute of Tea Research. 1974. Control of diseases and insects of tea. Includes scientific names (in
Latin) of insects. Agricultural Press, Peking. 166 pp. including
67 color plates.

Chinese Academy of Agricultural Sciences, Institute of Plant Protection.
1974. Combat wheat diseases and insects. Agricultural Press, Peking.
63 pp.

Cho Chi. 1972. Rice borers and their control. Shanghai People's Press,
Shanghai. 231 pp.

Chu Hung-fu. 1975. Pictorial guide to moths. Vol. 2 of pictorial guide to insects. Includes indexes to scientific names (in Latin) and Chinese names of insects. Scientific Press, Peking. 158 pp. plus 58 color plates.

Editorial Committee. 1972. Control of armyworm. Agricultural Press, Peking. 63 pp.

Editorial Committee. 1972. Pictures of Chinese crop diseases and insect pests. Vol. 2: Diseases and insects of wheat. Includes scientific names (in Latin) and Chinese names of insects and pathogens. Agricultural Press, Peking. 86 pp. including 42 color plates.

Editorial Committee. 1972. Pictures of Chinese crop diseases and insect pests. Vol. 4: Diseases and insects of cotton and hemp. Includes scientific names (in Latin) and Chinese names of insects and pathogens. Agricultural Press, Peking. 93 pp. including 45 color plates.

Editorial Committee. 1973. Color pictures of crop diseases and insect pests. Composite edition. Shanghai People's Press, Shanghai. No. 1, Rice (32 color plates); No. 2, Cotton (22 color plates); No. 3, Wheat, rape, and green manure crops (19 color plates); No. 4, Grain and soybean (20 color plates); No. 5, Vegetables (20 color plates); No. 6, Fruit trees (19 color plates); No. 7, Other crops (15 color plates).

Editorial Committee. 1973. Control of borers of corn, grain and soybean. Agricultural Press, Peking. 69 pp.

Editorial Committee. 1973. Handbook for control of diseases and insects of mulberry and silkworm. Agricultural Press, Peking. 198 pp. including 46 color plates.

Editorial Committee. 1973. How to grow mulberry and silkworm. Shanghai People's Press. 148 pp. plus four color plates.

Editorial Committee. 1973. Insect control and forecasting. Shanghai People's Press, Shanghai. 317 pp. plus 60 color plates.

Editorial Committee. 1973. Sugarbeet handbook. Includes scientific names (in Latin) of insects, pathogens, and plants. Inner Mongolia People's Press, Huhehaot. 287 pp. plus two color plates.

Editorial Committee. 1974. Control of diseases of wheat. Includes scientific names (in Latin) of pathogens. Agricultural Press, Peking. 196 pp.

Editorial Committee. 1974. Forest and fruit tree disease and insect control. Includes scientific names (in Latin) of pathogens and insects. Inner Mongolia People's Press, Huhehaot. 197 pp. plus 81 color plates.

*Editorial Committee. 1974. Handbook for plant protection workers. Composite edition. Includes indexes to scientific names (in Latin) and Chinese names of pathogens, poisonous plants, insects, and common pesticides. Shanghai People's Press, Shanghai. No. 1, Rice (66 pp. plus color plates); No. 2, Cotton (41 pp. plus color plates); No. 3, Wheat, rape, and green manure crops (39 pp. plus 19 color plates); No. 4, Grains, soybean (26 pp. plus 20 color plates); No. 5, Vegetables (33 pp. plus 20 color plates); No. 6, Fruit trees (32 pp. plus 19 color plates); No. 7, Other crops (20 pp. plus 15 color plates); No. 8, Common agricultural chemicals (86 pp.).

Editorial Committee. 1975. Agricultural microbiology. Shanghai People's Press, Shanghai. 241 pp.

*Editorial Committee. 1975. Control of soil insects. Agricultural
Press, Peking. 67 pp.

Editorial Committee. 1975. How to raise honey bees. Shanghai People's
Press, Shanghai. 212 pp.

Fei yo-chung. 1975. Questions and answers on agricultural chemicals.
Fuel Chemical Industry Press, Peking. 373 pp.

*Honan College of Agriculture, Entomology Teaching and Research Division.
1975. Pictorial guide to agricultural insects in Honan. Includes
indexes to scientific names (in Latin) and Chinese names of insects.
Honan College of Agriculture, Hsu-chang. 287 pp.

*Honan College of Agriculture, Entomology Teaching and Research Division.
1975. Pictorial guide to insects of fruit and forest trees in Honan.
Includes indexes to scientific names (in Latin) and Chinese names of
insects. Honan College of Agriculture, Hsu-chang. 379 pp.

Hopei Institute of Fruit Tree Research Revolutionary Committee. 1972.
Handbook of fruit tree cultivation. Agricultural Press, Peking.
722 pp. plus 24 color plates.

Hua County (Kwangtung Province) Revolutionary Committee. 1974. Selected
articles on experimentation in agricultural sciences by the masses.
Agricultural Press, Peking. 87 pp.

Hunan Province Revolutionary Committee, Agriculture and Forestry Bureau.
1973. Control of diseases and insect pests of cotton, hemp, sugar beet
and tobacco. Includes scientific names (in Latin) and Chinese names
of insects and pathogens. Hunan People's Press, Hunan. 89 pp. plus
43 color plates.

Hupei Province, Kwang kang District Revolutionary Committee, Agriculture
Bureau. 1973. Fifty questions on rice borer control. Hupei People's
Press, Hupei. 30 pp.

Hupei Province Revolutionary Committee, Agriculture and Forestry Bureau.
1972. Control of diseases and insects on fruit trees, vegetables
and specialty crops. Hupei People's Press, Hupei. 77 pp. plus 36
color plates.

Kai Cheh-lui and Hsien Ying-chi. 1965. Development in acarology. Scientific and Technical Press, Shanghai. 332 pp.

Kirin Provincial Institute of Forestry. 1972. Utilization of *Trichogramma*. Agricultural Press, Peking. 26 pp.

*Kou Chung-ling. 1966. Homoptera: Cicadellidae. Vol. 10 of Chinese
economic insects. Includes indexes to scientific names (in Latin) and
Chinese names of insects and host plants. Scientific Press, Peking.
170 pp.

Kwangchow Institute of Termite Control. 1972. Control of common termites. Kwangtung People's Press. 51 pp.

Kwangsi Academy of Agriculture. 1974. Control of rice leaf roller with
Trichogramma. Kwangsi People's Press, Kwangsi. 24 pp.

Kwangsi Institute of Forestry. 1974. Control of pine caterpillar.
Kwangsi People's Press, Kwangsi. 56 pp.

Kwangsi Province, Kang Autonomous District Revolutionary Committee,
Agriculture Bureau. 1973. Control methods of rice diseases and
insects. Kwangsi People's Press, Kwangsi. 130 pp. plus 50 color
plates.

Kwangtung College of Agriculture and Forestry, Department of Forestry, Research Division of Forest Diseases and Insects. 1975. Control of diseases and insects of lumber and bamboo. Includes scientific names (in Latin) of pathogens and insects. Kwangtung People's Press, Kwangtung. 111 pp.

Kwangtung Institute of Entomology and Biology Department, Chungshan University. 1973. Use of *Anastatus* to control lichee stink bug. Kwangtung People's Press, Kwangtung. 40 pp.

Kwangtung Provincial Academy of Agricultural Sciences. 1973. Handbook of control methods of major rice diseases and insects. Kwangtung People's Press, Kwangtung. 96 pp. plus 11 color plates.

Luh Pau-ling. 1974. Chinese Anophelinae identification handbook. 2nd ed. 34 pp. plus 40 plates. Scientific Press, Peking.

Nantung District Crop Disease and Insect Monitoring and Forecasting Station. 1971. Monitoring and forecasting of cotton pink bollworm. Shanghai People's Press, Shanghai. 26 pp.

Northwest Institute of Agriculture. 1956. Research and control of wheat midges. Includes scientific names (in Latin) of insects. Shensi People's Press, Shensi. 75 pp. plus two color plates.

Peking Institute of Agricultural Sciences. 1975. Apiculture. Agricultural Press, Peking. 94 pp. plus two color plates.

Shanghai Agricultural Chemical and Machinery Factory. 1966. How to use sprayers and dusters. Shanghai Scientific and Technical Press, Shanghai. 58 pp.

Shanghai Agriculture Bureau Revolutionary Committee Production Division. 1971. Development and control of cotton pink bollworm. Shanghai People's Press, Shanghai. 30 pp. plus two color plates.

Shansi Che County Nanhu Brigade Scientific Team and Sansi Che County Yuenping School of Agriculture. 1974. Activity pattern and control of mole crickets. Agricultural Press, Peking. 47 pp.

Shantung Peanut Research Institute. 1975. Diseases and insects of peanuts. Includes scientific names (in Latin) of plants, pathogens, and insects. Agricultural Press, Peking. 71 pp. plus 24 color plates.

Shantung Provincial Revolutionary Committee Agriculture Bureau. 1972. Control methods of cereal diseases and insects. Shangtung People's Press, Shangtung. 141 pp. plus 26 color plates.

Shangtung Provincial Revolutionary Committee, Agriculture Bureau. 1972. Control methods of diseases and insects on specialty crops and vegetables. Shangtung People's Press, Shangtung. 94 pp. plus 20 color plates.

Shensi Biological Resources Survey Team. 1974. Wax insects and wax production. Shensi People's Press, Shensi. 69 pp.

Sung Chi-ho. 1975. Experimental design and analysis for tests of agricultural chemicals. Scientific Press, Peking. 168 pp.

Szechuan Provincial Academy of Agricultural Sciences. 1974. Hybrid corn. Szechuan People's Press, Szechuan. 126 pp.

Szechuan Provincial Academy of Agriculture Institute of Agricultural Chemicals. 1973. Handbook of agricultural chemicals. Agricultural Press, Peking. 461 pp.

Szechuan Provincial Nan-yung School of Agriculture. 1964. Sericulture. Agricultural Press, Peking. 155 pp.

Ta-ku Chemical Engineering. 1975. Synthesis of BHC. Petroleum Chemical Industry Press, Peking. 213 pp.

Tang Chiu and Yao Kuo-jeh. 1957. Biological principles of controlling stored products insects with low temperature (translation of Russian). Scientific Press, Peking. 81 pp.

Tientsin Municipal Agriculture and Forestry Bureau, Extension Station. 1974. Control methods of diseases and insects of agricultural crops. Tientsin People's Press, Tientsin. 132 pp. plus 75 color plates.

*Tsai Pong-hwa and Chen Ning-shen. 1964. Isoptera: Termites. Vol. 8 of Chinese economic insects. Includes indexes to scientific names (in Latin) and Chinese names of insects. Scientific Press, Peking. 141 pp. plus eight plates.

Tsai Pong-hwa. 1973. Insect taxonomy. Vol. 2 of a 3-volume series. Includes indexes to scientific names (in Latin) and Chinese names of insects. Scientific Press, Peking. 303 pp. plus four color plates.

*Wu yeng-ju. 1965. Hymenoptera: Apoidea. Vol. 9 of Chinese economic insects. Includes indexes to scientific names (in Latin) and Chinese names of insects and plants. Scientific Press, Peking. 83 pp. plus seven color plates.

*Yang Wei-i. 1962. Hemiptera: Pentatomidae. Vol. 2 of Chinese economic insects. Includes indexes to scientific names (in Latin) and Chinese names of insects. Scientific Press, Peking. 138 pp. plus 10 plates.

Appendix E

GERM PLASM PRESENTED AND RECEIVED

I. Crop seeds presented to Chinese institutions:

A. Peking Institute of Zoology (to be distributed to appropriate scientists in Peking and elsewhere)

1. Wheat
   a. Courtesy of Dr. P. L. Adkisson (Texas A & M University):
      Texas R Line 2503-3-1. Hard red winter wheat germ plasm, semidwarf plant height, restored fertility to *Triticum timopheervi* cytoplasm
      TX 62A2712-6 Hard red winter wheat germ plasm, semi-dwarf height, pedigree Sturdy Stem, 391-56-D8/ Tascosa.
      TX 69A345-2 Hard red winter wheat germ plasm, semi-dwarf plant height, pedigree Short Wheat/Scout.
   b. Courtesy of Dr. Everett H. Everson (Michigan State University) through Dr. G. E. Guyer (Michigan State University):
      IONIA--Soft white winter wheat with W38 resistance to Hessian fly.
      TECUMSEH--Soft white winter wheat with W28 resistance to Hessian fly.
      4485 700747 CI 9321/2*Ge//Tecumseh--resistance to *Oulema melanopa* (cereal leaf beetle).
      4464 700742 CI 9321/2*Ge//Ionia--resistant to *Oulema melanopa* (cereal leaf beetle), races A and C of *Phytophaga destructor* (Hessian fly), and leaf rust.
      3373 680156 Yorkstar//Ge*6/P4217--carries the Ribero resistance to *Phytophaga destructor* (Hessian fly)-- resistant to all known races.
      3538 680328 Ge*4/Norin 10 Brevor//AC 4835/4*Ge--carries the Durum resistance (H5 Gene) to Hessian fly.
   c. Courtesy of Dr. R. E. Heiner (ARS, USDA) through Dr. H. C. Chiang (University of Minnesota):
      Chris (CI No. 13751) Awnless, medium height and maturity, moderately resistant to lodging.

212

Era (CI No. 13986) Awned, mid-season to late, semi-dwarf with high lodging resistance.

Ellar (CI No. 17289) Awnless, early, medium height, and lodging resistance.

Fletcher (CI No. 13985) Awned, mid-season to late, semidwarf with high lodging resistance.

Kitt (CI No. 17297) Awned, mid-season to late, semi-dwarf with high lodging resistance.

Manitou (CI No. 13775) Awnless, medium height and maturity, fair resistance to lodging.

Neepawa (CI No. 15073) Awnless, early, medium height, medium resistance to lodging.

Olaf (CI No. 15930) Awned, semidwarf, medium maturity with high resistance to lodging.

Polk (CI No. 13773) Awned, bronze chaff, medium height, medium resistance to lodging.

Waldron (CI No. 13958) Awnless, yellow chaff, early, medium height, very resistant to lodging.

2. Corn
   a. Courtesy of Dr. Jon Geadelmann (University of Minnesota) through H. C. Chiang:
      B52
      A619
      Erto (M) C4
      Eto (M) C6
   b. Courtesy of Dr. W. A. Russell (Iowa State University) through Dr. H. C. Chiang:
      CI 31A
      Oh 43
      B 49
      B 75

3. Cotton
   a. Courtesy of Dr. P. L. Adkisson:
      BRS-10 Glabrous, nectariless, Lukefahr
      TAMCOT-Sp-21
      TAMCOT-Sp-37
      HGE 12 High gossypol, hirsute, nectariless, Lukefahr

4. Sorghum
   a. Courtesy of Dr. P. L. Adkisson:
      TAM 428 FS'74 Male pollinator
      TAM 2566 L75CL4 Sorghum midge resistant R line
      TAM 2567 FS'75 Greenbug resistant R line
      A-399 FS'74 Male sterile female

5. Peanuts
 a. Courtesy of Dr. P. L. Adkisson:
   Florunner Drought resistant
   Starr Drought resistant
   Span Cross
   TAMNUT'74
   Spantex

B. Kirin Academy of Agricultural Sciences

1. Soybean
 a. Courtesy of Dr. J. W. Lambert (University of Minnesota)
   thorugh Dr. H. C. Chiang:
   Group 00 (Early)
    Ada
    Altona
    Norman
    Portage
    M-65-217
   Group 0 (Medium early)
    Clay
    Evans
    Merit
    Swift
    Wilkin
    M-65-94
    M-65-295
   Group I (Medium late)
    Anoka
    Chip 64
    Hark
    Harlon (OX 643)
    Hodgson
    Steele
    M-65-115
    M-65-442
   Group II (Late)
    Amsoy 71
    Beeson
    Corsay
    Rampage
    Wells
 b. Courtesy of Dr. W. H. Luckmann (Illinois Natural History Survey) through Dr. H. C. Chiang:
   PI 227, 687
   PI 171, 451

2. Corn
   a. Courtesy of Jon Geadelmann through Dr. H. C. Chiang:
      B52
      A619
      Erto (M) C4
      Eto (M) C6
   b. Courtesy of Dr. A. Russell through Dr. H. C. Chiang:
      CI 31A
      Oh 43
      B 49
      B 75

C. Shanghai, Agricultural Academy of Sciences

   1. Wheat (courtesy of Dr. P. L. Adkisson):
      Hard red wheat TX 65A1503-1
      Hard red wheat TX 69A525-1
      Hard red wheat TX 62A2782-4-2

   2. Cotton (courtesy of Dr. P. L. Adkisson):
      HGE 12 High gossypol, hirsute, nectariless
      BRS 10 Glabrous, nectariless
      1209-619-7 Semidwarf
      1073-166 Nectariless
      407-26 Glabrous
      407-79-R2 Frego bract
      IX6-56 Semidwarf
      TAMCOT-Sp-21
      TAMCOT-Sp-32
      HG-BR-8 High gossypol

   3. Sorghum (courtesy of Dr. P. L. Adkisson):
      A399 FS'74
      TAM 2567 FS'75
      TAM 2566 L75 CL4
      TAM 428 FS'74

   4. Peanuts (courtesy of Dr. P. L. Adkisson):
      Florunner
      Span Cross
      Spantex
      Starr

D. Hunan Provincial Institute of Plant Protection

   1. Sorghum (courtesy of Dr. P. L. Adkisson):
      TAM 2567 FS'75
      TAM 2566L75CL4
      A 399 FS'74
      TAM 428 FS'74

2. Peanuts (courtesy of Dr. P. L. Adkisson):
    Starr
    Stantex
    TAMNUT 74
    Florunner
    Span Cross

II. Seeds received from Chinese institutions:

A. Kirin Provincial Institute of Agricultural Sciences

    1. Soybean
        Kirin No. 3
        Kirin No. 4
        Kirin No. 8
        Kirin No. 9
        Kirin No. 10
        Kirin No. 11
        Tsiao-fong No. 1
        Chi-ti No. 3
        Chi-ti No. 5
        Small White
        Early maturing dwarf

    2. Corn inbreds
        Hua 94
        84-74
        Ying 55
        Ying 46
        Tieh 84
        Tieh 133
        Chi 69

    3. Sorghum
        Whu No. 22 (grey)
        Tsuo-sterile line
        Tsuo-fertile line

    4. Wheat
        Fong-chiang No. 2
        Liao-chung No. 6

B. Shanghai, Agricultural Academy

    1. Cotton
        No. 613

    2. Rice
        South early dwarf No. 1
        Shanghai late No. 19

3. Wheat
    Spring wheat No. 757
    Shanghai wheat

C. Hunan Provincial Institute of Plant Protection

1. Rice
    Hunan early dwarf No. 4
    Hunan early dwarf No. 8

Appendix F

CONVERSION TABLE

For most Americans, China's system of weights and measures may seem confusing since the metric system is used in all international transactions and a Chinese system for domestic exchanges. Following is a partial list of equivalents:

1 liang (tael)	=	2.5 ounces	=	50 grams
1 jin (catty)	=	1.102 pounds	=	0.5 kilogram
1 gungjin	=	2.204 pounds	=	1 kilogram
1 dan (picul)	=	110.23 pounds	=	50 kilograms
1 mu	=	0.1647 acre	=	0.066 hectare
1 li	=	0.35 mile	=	0.56 kilometer
1 gungli	=	0.621 mile	=	1 kilometer

The Chinese are phasing out the *jin* and *li* units and making the *gungjin* (kilogram) and *gungli* (kilometer) standard.